Günther Wolfram (Ed.)

Genetic and Therapeutic Aspects of Lipid and Purine Metabolism

With 59 Figures and 33 Tables

Springer-Verlag Berlin Heidelberg New York
London Paris Tokyo

Professor Dr. med. GÜNTHER WOLFRAM

Medizinische Poliklinik der Universität
Pettenkoferstrasse 8a, 8000 München 2
und
Institut für Ernährungswissenschaft
in Weihenstephan
Technische Universität München
8050 Freising, Federal Republic of Germany

ISBN 3-540-50408-7 Springer-Verlag Berlin Heidelberg New York
ISBN 0-387-50408-7 Springer-Verlag New York Berlin Heidelberg

© Springer-Verlag Berlin Heidelberg 1989
Printed in Germany

The use of registered names, trademarks, etc. in this publication does not imply, even in the at
a specific statement, that such names are exempt from the relevant protective laws and regulatic
therefore free for general use.

Product Liability: The publisher can give no guarantee for information about drug dosage and application thereof contained in this book. In every individual case the respective user must check its accuracy by consulting other pharmaceutical literature.

Typesetting, printing and bookbinding: Brühlsche Universitätsdruckerei, Giessen
2121/3145-543210 – Printed on acid-free paper

In honour of

Nepomuk Zöllner

on the occasion of his 65th birthday

Preface

This publication of a symposium held on 24th and 25th of June 1988 in Munich is dedicated to Nepomuk Zöllner on the occasion of his 65th birthday, expressing the best wishes of the authors.

Nepomuk Zöllner was born in the northern part of Bavaria. While a medical student in Munich he was called up to military duty in the last year of World War II. After achieving excellent results on his examinations he served as a physician in several hospitals in Munich and soon became interested in inborn errors of metabolism. In order to receive the best education possible at that time he joined the group of S. J. Thannhauser, the famous German emigrant, in Boston where he worked from 1951 to 1953. There he was able to continue his studies on lipids, which he had started in 1948, and begin with his studies on purine metabolism. While working in Thannhauser's laboratory he learned to think and act according to the strict laws of natural science.

After returning to Europe he concentrated his scientific work on purines and lipids. Gout and hyperlipoproteinemias from the genetic to the therapeutic aspects remained the dominating topics of his research activities. In spite of many obligations as head of the Medical Polyclinic of the University of Munich and many activities at the university and in scientific societies Zöllner's foremost intent continued to be the development and progress of natural science in medicine. In addition to his research activities in gastroenterology, cardiology, pneumology, infectious diseases, and clinical nutrition he very successfully achieved advances in diagnosis and more specific and effective therapies of inborn errors of metabolism like familial hyperlipidemias and gout.

Nepomuk Zöllner was one of the European scientists who founded the European Atherosclerosis Group in 1964 to promote meetings of European scientists to exchange ideas in the various disciplines of atherosclerosis research.

This book contains an integrated and up-to-date review of fascinating research work and stimulating approaches to practical medicine. All important topics are addressed. Starting with the genetic aspects of lipid metabolism, i.e., DNA polymorphisms, genetic aspects of LDL receptors, scavenger receptor in human liver, LDL metabolism or HDL structure, the book continues with purine metabolism, i.e., human HPRT deficiency, myoadenylate deaminase deficiency, purine excretion, and nephrolithiasis. Discussions of the clinical aspects of hyperlipidemias from prognosis and risk factors to modern therapy by diet, drugs, and plasmapheresis follow as do papers on the clinical aspects of gout, including diet and drug therapy. With the development of antiviral compounds against HIV and Herpesvirus, awarded a Nobel Prize in 1988 (G. B. Elion and G. H. Hitchings), Nepomuk Zöllner's most recent work on AIDS is also referred to.

This book and other papers published in 1988 in *Klinische Wochenschrift* (66:85–211) are dedicated to Nepomuk Zöllner and reflect his lifelong effort to improve our understanding of human biology and medicine so that we are better equipped to diagnose and treat metabolic diseases. These contributions offer a critical synthesis of present knowledge in various subjects by the most experienced experts among his pupils and friends.

Ad multos annos! G. WOLFRAM

Table of Contents

Therapeutic Aspects

Hyperlipidemia

Disturbances of Purine Metabolism

List of Contributors

Dr. E. BINCZEK
Institut für Physiologische Chemie der Universität, Josef-Stelzmann-Strasse 9,
5000 Köln 41, Federal Republic of Germany

Priv.-Doz. Dr. H. BORBERG
Medizinische Universitätsklinik, Josef-Stelzmann-Strasse 9, 5000 Köln 41,
Federal Republic of Germany

Dr. B. L. DAVIDSON
Department of Internal Medicine, University of Michigan, Medical Center,
Ann Arbor, MI 48109, USA

Prof. Dr. R. GREGER
Physiologisches Institut, Albert-Ludwigs-Universität,
Hermann-Herder-Strasse 7,
7800 Freiburg, Federal Republic of Germany

Prof. Dr. H. GRETEN
Medizinische Klinik der Universität Hamburg, Martinistrasse 52,
2000 Hamburg 20, Federal Republic of Germany

Prof. Dr. W. GRÖBNER
Kreiskrankenhaus Balingen, Tübinger-Strasse 30, 7460 Balingen,
Federal Republic of Germany

Dr. K. HARDERS-SPENGEL
Medizinische Poliklinik der Universität München, Pettenkoferstrasse 8a,
8000 München 2, Federal Republic of Germany

Dr. A. HAASE
Institut für Physiologische Chemie der Universität, Josef-Stelzmann-Strasse 9,
5000 Köln 41, Federal Republic of Germany

Prof. Dr. E. W. HOLMES
Department of Medicine, Division of Metabolism,
Endocrinology and Genetics,
Duke University Medical Center, Durham, NC 27710, USA

Dr. C. HOLTFRETER
Institut für Physiologische Chemie der Universität, Josef-Stelzmann-Strasse 9,
5000 Köln 41, Federal Republic of Germany

Dr. S. HUMPHRIES
The Charing Cross Sunley Research Centre, Lurgan Avenue,
Hammersmith, London W6 8LW, Great Britain

Prof. Dr. CH. KELLER
Medizinische Poliklinik der Universität München, Pettenkoferstrasse 8a,
8000 München 2, Federal Republic of Germany

Prof. Dr. W. N. KELLEY
Department of Internal Medicine, University of Michigan, Medical Center,
Ann Arbor, MI 48109, USA

Prof. Dr. K. OETTE
Institut für Klinische Chemie der Universität, Josef-Stelzmann-Strasse 9,
5000 Köln 41, Federal Republic of Germany

Dr. C. J. PACKARD
Department of Pathological Biochemistry, University of Glasgow,
Royal Infirmary, Glasgow G4 OSF, Great Britain

Dr. T. D. PALELLA
Department of Internal Medicine, University of Michigan, Medical Center,
Ann Arbor, MI 48109, USA

Dr. K. L. POWELL
Department of Biochemical Virology, Wellcome Research Laboratories,
Langley Court, Beckenham, Kent BR3 3BS, Great Britain

Dr. G. RAUH
Medizinische Poliklinik der Universität München, Pettenkoferstrasse 8a,
8000 München 2, Federal Republic of Germany

Dr. R. L. SABINA
Department of Medicine, Division of Metabolism,
Endocrinology and Genetics,
Duke University Medical Center, Durham, NC 27710, USA

Prof. Dr. G. SCHLIERF
Klinisches Institut für Herzinfarktforschung, Bergheimer Strasse 58,
6900 Heidelberg, Federal Republic of Germany

Priv.-Doz. Dr. G. Schuler
Kardiologie, Klinisches Institut für Herzinfarktforschung,
Bergheimer Strasse 58, 6900 Heidelberg, Federal Republic of Germany

Dr. H. Schuster
Medizinische Poliklinik der Universität München, Pettenkoferstrasse 8a,
8000 München 2, Federal Republic of Germany

Prof. Dr. P. Schwandt
Medizinische Klinik II der Universität, Klinikum Grosshadern,
Marchioninistrasse 15, 8000 München 70, Federal Republic of Germany

Prof. Dr. D. Seidel
Klinische Chemie, Klinikum der Universität, Robert-Koch-Strasse 40,
3400 Göttingen, Federal Republic of Germany

Prof. Dr. J. Shepherd
Department of Pathological Biochemistry, University of Glasgow,
Royal Infirmary, Glasgow G4 OSF, Great Britain

Dr. H. A. Simmonds
Guy's Hospital, Purine Laboratory, Guy's Tower (17th&18th Floors),
London Bridge SE1 9RT, Great Britain

Dr. B. Stiefenhofer
Medizinische Poliklinik der Universität München, Pettenkoferstrasse 8a,
8000 München 2, Federal Republic of Germany

Prof. Dr. W. Stoffel
Institut für Physiologische Chemie der Universität, Josef-Stelzmann-Strasse 9,
5000 Köln 41, Federal Republic of Germany

Dr. T. C. Südhof
Department of Molecular Genetics, University of Texas, Health Science
Center at Dallas, 5323 Harry Hines Blvd., Dallas, TX 75235, USA

Dr. R. Taylor
The Charing Cross Sunley Research Centre, Lurgan Avenue,
Hammersmith, London W6 8 LW, Great Britain

Prof. Dr. G. Wolfram
Medizinische Poliklinik der Universität München, Pettenkoferstrasse 8a,
8000 München 2, Federal Republic of Germany

Prof. Dr. H. F. Woods
University of Sheffield, Royal Hallamshire Hospital, Glossop Road,
Sheffield S10 2JF, Great Britain

Genetic Aspects

Lipid Metabolism

DNA Polymorphisms in the Diagnosis of Familial Hypercholesterolaemia

H. Schuster, B. Stiefenhofer, G. Rauh, R. Taylor, and S. Humphries

Introduction

Familial hypercholesterolaemia (FH) is a common, inherited disease with a frequency of about 1 in 500. The disorder is characterized clinically by elevation in the concentration of low-density lipoprotein (LDL) cholesterol in blood, tendon xanthomata, an increased risk of myocardial infarction and, genetically, by autosomal dominant inheritance [1–3]. FH contributes significantly to the number of individuals suffering from premature coronary artery disease in Western society [4]. If individuals with FH were identified before they develop symptomatic disease, they could be treated prophylactically to reduce the future risk of myocardial infarction [5]. To date, however, the disease cannot always be diagnosed unequivocally in early childhood on clinical signs only.

The identification of potentially affected individuals in families has, in the past, relied upon the measurement of serum and LDL cholesterol levels determined from peripheral blood and cord blood [6–8]. However, serum cholesterol levels do not always allow unequivocal diagnosis of FH, especially when determined from cord blood [6]. The values obtained in this way sometimes lie within the normal reference ranges and may not rise to levels at which definite diagnosis of FH is possible until later in life [9]. This problem could be circumvented by carrying out serial blood cholesterol determinations, but this would be inconvenient and expensive. The use of a direct genetic test for FH could provide an unequivocal diagnosis.

Diagnosis has also been attempted by culturing fibrolasts from a skin biopsy in order to directly measure the binding of labelled LDL to the cell LDL receptors [10] or by examining functional LDL receptors on cultured lymphocytes [11]. These methods are technically difficult and, again, overlapping values can be obtained from normal individuals and individuals with heterozygous FH [12]. It is therefore possible that FH would not be

3

diagnosed by these methods in children until coronary artery disease was already emerging as a result of rising cholesterol levels.

The cloning of the human LDL receptor cDNA has now made it possible to analyse the defects at the DNA level [13]. Up to now at least 20 different mutations of the LDL receptor gene have been reported. Two categories of gene defects have been shown to cause FH. The first category includes gross structural alterations to the gene such as deletions or insertions [14]. The second category includes defects which arise due to a single base pair exchange in the gene either abolishing the synthesis of the receptor or resulting in the protein being inactive [15]. A number of restriction fragment length polymorphisms (RFLPs) of the LDL receptor gene have been reported which can be used for presymptomatic diagnosis in families [16].

DNA Rearrangements

In order to look for gross alterations in the LDL receptor gene, DNA samples were studied from 60 unrelated FH patients from London (55 heterozygous and 5 homozygous, 65 defective genes) and 100 unrelated FH patients from Munich (97 heterozygous and 3 homozygous, 103 defective genes). The restriction enzymes used were BglII and XbaI. Two different probes for the LDL receptor cDNA were used, one for the 5' part of the

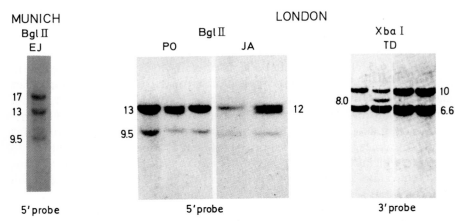

Fig. 1. Genomic Southern blot analysis of the LDL receptor gene in FH patients from London and Munich. 5 µg of DNA from each patient was digested with XbaI or BglII, size fractionated on a agarose gel, transferred to filter membranes and hybridized with the radioactively labelled LDL receptor probe as described [14]. The 3' probe consists of a 1.9-kb BamHI fragment (base pairs 1573–3486) of the 3' part of the LDL receptor cDNA. The 5' probe consists of a 1.8-kb HindII/BglII fragment from plasmid pLDLR-3 and covers nucleotides 1–1700

gene (*Bgl*II digest), and one for the 3′ part of the gene (*Xba*I digest; Fig. 1).

In four of the London patients, the probes detected both gene fragments of the expected size and an additional smaller fragment (Fig. 1). In each patient the smaller fragments are the result of a partial deletion of one of the alleles of the LDL receptor gene. A detailed analysis of the defect in the patient TD was carried out. The deletion removes about 5 kb of DNA, including exons 13 and 14 [14]. The defective gene was isolated and the sequence of the DNA in the region determined [17]. Analysis indicates that the deletion occurred as a result of a recombination event between two repeat "Alu" sequences in the introns of the gene. The sequence analysis predicts that the defective gene should produce a truncated protein that lacks the COOH-terminal 230 amino acids. This region of the protein is involved in anchoring the protein in the cell membrane and if transported to the membrane, the protein would not be retained in the coated pit. However, using Western blotting a truncated LDL receptor protein could not be detected on the surface of the patient's fibroblasts or in cytoplasmic extracts of the cells. It thus appears that the truncated protein may be rapidly turned over within the cell.

A detailed map of the deletions detected with the 5′ probe suggests that the deletions are 1–2 kb in length [18] and may remove regions of the gene coding for part of the LDL-binding domain or part of the epidermal growth factor "stem" region. Identification of the relatives of the patients who have inherited the defective allele of the LDL receptor gene will thus be possible in all these families, based on direct DNA analysis, as demonstrated in the family of TD [14].

In the Munich patients only one individual was identified with a different pattern of fragments detected with the *Bgl*II digest and the 5′ probe (Fig. 1). Preliminary analysis indicates that an insertion of 3–4 kb in length has occurred as a result of a recombinant event between the Alu sequences in intron 4 and intron 6 and thus the defect of the protein can be predicted to be in the binding site for LDL [19]. Similar deletions involving a recombination event between Alu sequences have been reported by several other workers. The frequency of these deletions in most populations studied is between 1% and 4%. However, in FH patients in Canada of French ancestry, a similar deletion occurs in about 60% of patients [20], and a different deletion occurs in about 30% of patients with FH in Finland [21]. This high frequency is probably the result of a founder gene effect.

Common DNA Polymorphisms: Diagnosis in Families

Although our screening method would not identify small deletions or insertions, the results suggest that the majority of defects in the LDL recep-

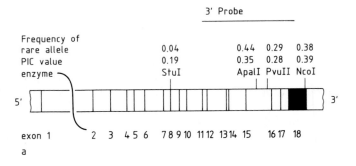

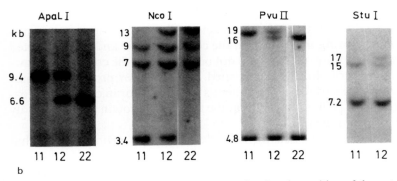

Fig. 2. a Map of the human LDL receptor gene showing the position of the restriction sites of the *Stu*I, *Apa*II, *Pvu*II and *Nco*I RFLPs. The 3' probe used to detect the polymorphic fragments spans from exon 11 to exon 18 (base pairs 1573–3486). **b** Autoradiogramm showing the polymorphic bands of the *Apl*I, *Nco*I, *Pvu*II and *Stu*I polymorphism detected with the 3' 1.9-kb *Bam*HI fragment of the human LDL receptor cDNA. Fragment sizes in kilobases

tor gene are due to point mutations rather than gross alterations in the gene. For the relatives of individuals with such defects, presymptomatic diagnosis can be carried out based on genetic linkage analysis using RFLPs.

To date common RFLPs of the LDL receptor gene have been detected using at least twelve enzymes [16, 22–29]. For some enzymes there are several different alleles, and taken together, these RFLPs should be informative in more than 90% of families. The relative usefulness of these different RFLPs depends both on their frequency in the population and on whether there is linkage equilibrium (random association) of the alleles of the two RFLPs with each other. We have examined this for the RFLPs detected with *Stu*I, *Apa*II, *Pvu*II and *Nco*I (Fig. 2a) in a sample of 69 normal individuals from London. In this sample we detected significant linkage disequilibrium between all the polymorphic sites [30]. Over such a small genetic distance the degree of linkage disequilibrium is not related primarily to the physical distance between the varying genetic markers, but rather

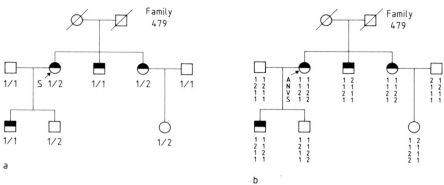

Fig. 3. a Pedigree of family no. 479. Linkage analysis between FH phenotype and the *Stu*I polymorphisms of the LDL receptor gene. The *S1* allele of the proposita cosegregates with the FH phenotype in this particular family. Therefore, diagnosis of FH can carried out in children of the proposita and her sister. Diagnosis of FH could not carried out in the brother's family, because we cannot distinguish the normal *S1* from the defective *S1* allele. **b** Pedigree of FH family no. 479. Linkage analysis between FH phenotype and the *Apa*II, *Nco*I, *Pvu*II and *Stu*I polymorphism of the LDL receptor gene. In the brother of the index case, the haplotype of the normal allele can be determined as *S1 A2 V1 N1* and the haplotype of the defect allele as *S1 A1 V2 N1*. Therefore, diagnosis of FH could be carried out in his children using the haplotype information

is due to the evolutionary history of the four polymorphisms. However, data are consistent with the physical map of the location of the polymorphisms, with the *Stu*I varying site being in the 5′ region of the gene and the *Apa*II, *Pvu*II and *Nco*I sites being closer together in the 3′ region of the gene (Fig. 2 a).

It is possible to get a standardised estimate of the relative usefulness of an RFLP or RFLPs taken together by calculating the polymorphism information content (PIC) value for the polymorphisms [31]. For two polymorphisms in linkage equilibrium the combined value is slightly less than the sum of the separate PIC values, whereas for RFLPs in strong linkage disequilibrium the combined PIC value may be only slightly greater than that for one RFLP alone. For the *Stu*I, *Apa*II, *Pvu*II and *Nco*I RFLPs this combined PIC value is >0.80. This suggests that using these four RFLPs alone will be informative for genetic studies and for early detection of FH in over three-quarters of families.

In the Lipid Clinic in Munich the families of 40 patients with FH and more than four living relatives have been studied using these four RFLPs. In 36 of these families (90%), the inheritance of the LDL receptor gene could be followed unambiguously using information from a combination of these RFLPs [32]. This confirms the described calculation of the usefulness.

Figure 3 a shows a family from Munich in which early diagnosis of FH was made using the RFLP detected with *Stu*I. For any individual with FH, there is no way of knowing a priori whether the defective LDL receptor gene is on the chromosome with the cutting site for *Stu*I ("*S1*" allele) or without the cutting site ("*S2*" allele). This has to be deduced by inspection of the pedigree and determining which chromosome is inherited by the affected and unaffected family members.

In this family, the proposita is hypercholesterolaemic and has the genotype *S1 S2*. Her brothers and sisters, who are both hypercholesterolemic, have the genotype *S1 S1* and *S1 S2*. All brothers and sisters share only the *S1* allele and must therefore have the defective gene on a *S1* allele carrying the cutting site. In this family the defective LDL receptor gene is therefore cosegregating with the *S1* allele of the DNA polymorphism. The children in this family who have inherited the defective LDL-receptor gene can be identified before they develop high serum lipid levels.

The husband of the proposita is normal and has the genotype *S1 S1*. The older son has inherited a normal *S1* allele of the LDL receptor gene from his father and the defective *S1* allele from his mother. The younger son has inherited a normal *S1* allele from his father also and the normal *S2* allele from his mother. The niece of the proposita has inherited a normal *S1* allele from her father and the normal *S2* allele from her mother.

However, diagnosis cannot carried out in the family of the proposita's brother, since the normal and the defective *S1* alleles cannot be distinguished. In order to be able to do this, it will be necessary to use other RFLPs of the gene. Information from several RFLPs can be combined to distinguish the haplotype of the defective and the normal alleles and to follow the inheritance of the genes in this family. In Fig. 3 b information from four RFLPs has been used to determine the defective and normal allele in the proposita's brother. The inheritance of the defective LDL receptor gene can now be determined unambigously.

Diagnosis in the General Population

It would be useful to develop a rapid DNA test for FH so that any individual detected with hyperlipidaemia could be screened for defects in the LDL receptor gene without resorting to family studies. At the present time this is not technically possible, and current evidence indicates that this is unlikely to be feasible in the near future. Information about what specific mutations are causing FH in a patient may be useful in developing therapeutic strategies if, for example, patients with a particular defect respond best to a certain drug or have a particular prognosis.

There are several methods that have been developed for detecting single base changes in genes. It is easiest to detect if the mutation creates or destroys the cleavage site for a particular restriction enzyme. This will result in a "mutation-specific" DNA polymorphism of the LDL receptor gene that can be detected using Southern blotting. For example, if the base pair change that causes the mutation creates a new cutting site for the enzyme, the normal gene will show only one fragment and the defective gene will be cleaved and show two fragments. To date, the only situation where this approach has been useful is the mutation that commonly causes FH in Lebanon which can be detected with a cDNA probe and the enzyme *Hin*fI [15].

The second method to detect a single base pair change uses a pair of oligonucleotides, one specific for the normal gene sequence and the other specific for the mutated sequence. This "oligonucleotide melting" technique allows the identification of this same mutation in other patients. This approach has been successfully applied to the detection of thalassaemias [33], sickle cell anaemia [34] and variants of the apolipoprotein E (apo E) [35]. In combination with gene amplification by the polymerase chain reaction [36] this will be a powerful tool in the future.

However, the problem is that many mutations that cause FH have been reported – at least 20 have been documented and there are probably many more. This means that to detect all known mutations, a battery of several dozen specific oligonucleotides would have to be used. At best this approach would only be able to exclude all known mutations as being involved in causing FH in the individual being tested. The problem would be simplified in situations where one or two mutations make up a significant proportion of all FH defects in a particular population and oligonucleotide tests could be developed for these mutations.

There are several examples known where the majority of FH patients in an area have the same mutation in the LDL receptor gene, presumably as a result of a founder effect with geographical or cultural isolation. This is seen in the Lebanese and Syrian population where, as mentioned above, the majority of patients with FH have a mutation that creates a site for the restriction enzyme *Hin*fI. Also, 60% of French Canadian patients with FH have a common deletion of part of the LDL receptor gene and 30% of patients from Finland share another defect. A fourth example is in the Afrikaner population in South Africa where FH is very common. In one small study 19 out of 20 FH genes examined wre on a chromosome defined by the *Stu*I and *Pvu*II haplotype *S1 V1* [37]. This happened because a specific mutation causing FH must have occurred on a chromosome with a particular haplotype, and due to a founder effect, all descendents from this individual with FH will have the same haplotype. Once the specific mutation causing FH in these patients is known, a test can be developed using the oligonucleatide melting technique. This might permit the detection of more than 90% of patients with FH in the Afrikaner population. It is thus of interest to determine the haplotype distribution of the defective LDL receptor genes in a defined population.

Tabelle 1. Four RFLP haplotypes of normal and defective LDL receptor gene alleles determined by linkage analysis in 36 heterozygous and three homozygous FH patients

Haplotype				Defective (n = 39)	Normal (n = 90)
Stu I	Apal I	Nco I	Pvu II		
+	+	+	−	21	29
+	−	−	−	8	24
+	−	+	+	7	20
+	+	−	−	1	7
+	−	−	+	1	1
−	−	+	+	1	2
+	−	+	−	0	3
+	+	+	+	0	2
−	−	−	+	0	2
−	−	−	−	0	1

+, Presence of restriction site;
−, absence of restriction site.

In families from Munich the four-RFLP haplotypes of 39 defective and 90 normal alleles has been determined by linkage analysis (Table 1) [32]. With four RFLPs there are 16 possible haplotypes, but because of the small sample size and linkage disequilibrium between the polymorphic sites, only ten different haplotypes were observed in the normal sample and six in the FH sample. This suggests that there are at least six different mutations causing FH in this sample, although this is a minimum estimate. It is likely that several different independent mutations have occurred on, for example, the common $S + A + N + V$-haplotype, which is therefore heterogenous. One way of pursuing this approach is to use other RFLPs at the 5′ end of the gene to try to distinguish different subgroups within this haplotype that may be associated with only one mutation.

Conclusions

There are now enough common DNA polymorphisms of the LDL receptor gene to be able to carry out family studies, where appropriate, for unequivocal diagnosis of FH in childhood.

The challenge for the future is to develop methods of detecting defects causing FH in the population without having to resort to family studies. At the present time this appears technically difficult. However, in certain geographical areas or countries it should be possible to develop tests that may identify a large proportion of but not all patients with FH. Coupled with family studies for the remainder of the patients this may be a useful approach.

Once identified, affected individuals could be treated prophylactically and encouraged from an early age to keep a low-fat diet, get enough physical exercise and avoid smoking. The recent results of the Primary Coronary Prevention study offer hope that early therapeutic intervention will effectivly lower the risk of developing premature atherosclerosis.

References

1. Thannhauser SJ, Magendantz H (1938) The different clinical groups of xantomatous dieseases; a clinical physiological study of 22 cases. Ann Int Med 11:1622–1746
2. Müller C (1938) Xanthomata, hypercholesterolaemia, angina pectoris. Acta Med Scand [Suppl] 89:75–84
3. Fredrickson DS, Goldstein JL, Brown MS (1978) The familial hyperlipoproteinemias. In: Stanbury JB, Wyngaarden JB, Fredrickson DS (eds) The basis of inherited disease. McGraw-Hill, New York, pp 604–655
4. Goldstein JL, Hazzard WR, Schrott HG, Bierman EL, Motulsky AG (1973) Hyperlipidemia in coronary heart disease: Part I. Lipid levels in 500 survivors of myocardial infarction. J Clin Invest 52:1533–1543
5. Lipid Research Clinics Program (1984) The lipid research clinics coronary primary prevention trial results: II. The relationship of reduction in incidence of coronary heart disease to cholesterol lowering. JAMA 251:356–374
6. Glueck CJ, Heckman F, Schoenfeld M et al. (1971) Neonatal familial type II hyperlipoproteinaemia: cord blood cholesterol in 1800 births. Metabolism 20:597–608
7. Kwiterovich PO, Levy RI, Frederickson DS (1973) Diagnosis of familial type-II hyperlipoproteinaemia. Lancet I:118–121
8. Leonard JV, Whitelaw AGL, Wolff OH, Lloyd JK, Slack J (1977) Diagnosing familial hypercholesterolemia in children by measuring serum cholesterol. Br Med J I:1566–1568
9. Kessling AM, Seed M, Shennen N, Niththyanathan R, Wynn V (1988) Rising cholesterol levels in children at risk of familial hypercholesterolemia. (In preparation)
10. Spengel F, Harders-Spengel K, Keller C, Wieczorek A, Wolfram G, Zöllner N (1982) Use of fibroblast culture to diagnose and genotype familial hypercholesterolemia. Ann Nutr Metab 26:240–247
11. Cuthbert JA, East CA, Bilheimer DW, Lipsky PE (1986) Detection of familial hypercholesterolemia by assaying functional low-density-lipoprotein receptors on lymphocytes. N Engl J Med 314:879–883
12. Keller C, Spengel F, Wieczorek A, Wolfram G, Zöllner N (1981) Familial hypercholesterolaemia: a family with divergence of clinical phenotype and biochemical genotype based on fibroblast studies. Ann Nutr Metab 25:79–84
13. Yamamoto T, Davis LG, Brown MS, Schneider WJ, Casey MI, Goldstein JL, Russell DW (1984) The human LDL-receptor: a cysteine-rich protein with multiple Alu sequences in its mRNA. Cell 39:27–38
14. Horsthemke B, Kessling AM, Seed M, Wynn V, Williamson R, Humphries SE (1985) Identification of a deletion in the low density lipoprotein (LDL) receptor gene in a patient with familial hypercholesterolaemia. Hum Genet 71:75–78
15. Lehrman MA, Schneider WJ, Brown MS, Davis CG, Elhammer A, Russell DW, Goldstein JL (1987) The Lebanese allele at the low density lipoprotein receptor locus. Nonsense mutation produces truncated receptor that is retained in endoplasmic reticulum. J Biol Chem 262:401–410
16. Humphries SE, Kessling AM, Horsthemke B, Donald JA, Seed M, Jowett NI, Holm M, Galton DJ, Wynn V, Williamson R (1985) A common DNA polymorphism of the low density lipoprotein (LDL) receptor gene and its use in diagnosis. Lancet I:1003–1005

17. Horsthemke B, Beisiegel U, Dunning A, Williamson R, Humphries S (1987) Non-homologous crossing over between two alu-repetitive DNA sequences in the LDL-receptor gene: a possible mechanism for a novel mutation in a patient with familial hypercholesterolaemia. Eur J Biochem 24:144–147
18. Horsthemke B, Dunning A, Humphries S (1987) Identification of deletions in the human low density lipoprotein receptor gene. Med Genet 24:144–147
19. Schuster H, Stiefenhofer B, Keller Ch, Wolfram G, Ley J, Zöllner N (1988) Gross alterations in the LDL-receptor gene in FH patients from Munich. (In preparation)
20. Hobbs HH, Brown MS, Russell DW, Davigon J, Goldstein JL (1987) Deletion in LDL receptor gene occurs in majority of French Canadians with FH. N Engl J Med 317:734–737
21. Aalto-Setälä K, Gylling H, Miettinen T, Kontula K (1988) Identification of a deletion in the LDL-receptor gene: a Finnish type of mutation. FEBS Lett 230:31–34
22. Leitersdorf E, Hobbs HH (1987) Human LDL receptor gene: two ApaII RFLPs. Nucleic Acid Res 15,6:2782
23. Hobbs HH, Esser V, Russel DW (1987) AvaII polymorphism in the human LDL receptor gene. Nucleic Acid Res 15:379
24. Steyn LT, Pretorius A, Brink PA, Bester AJ (1987) RFLP for the human LDL receptor gene (LDLR): BstEII. Nucleic Acid Res 15,11:4702
25. Geisel J, Weisshaar B, Oette K, Mechtel M, Doerfler W (1987) Double MspI RFLP in the human LDL receptor gene. Nucleic Acid Res 15,9:3943
26. Kotze MJ, Retief AE, Brink PA, Weich HFH (1986) A DNA polymorphism in the human low-density lipoprotein receptor gene. S Afr Med J 70:77–79
27. Funke H, Klug J, Frossard P, Coleman R, Assman G (1986) PstI RFLP close to the LDL receptor gene. Nucleic Acid Res 14:7820
28. Kotze MJ, Langenhoven E, Dietzsch E, Retief AE (1987) A RFLP associated with the low-density lipoprotein receptor gene (LDLR). Nucleic Acid Res 15,1:376
29. Hobbs HH, Leitersdorf E, Goldstein JL, Brown MS, Russell DW (1988) Multiple crm-mutations in familial hypercholesterolaemia. Evidence for 13 alleles, including four deletions. J Clin Invest 81:909–917
30. Taylor R, Jeenah M, Seed M, Humphries SE (1987) Four DNA polymorphisms of the LDL-receptor gene: their genetic relationships and use in diagnosis of familial hypercholesterolaemia. J Med Genet (In press)
31. Botstein D, White RL, Skolnick M, Davis RW (1980) Construction of a genetic linkage map using restriction fragment length polymorphisms. Am J Hum Genet 32:314–331
32. Schuster H, Stiefenhofer B, Wolfram G, Keller Ch, Humphries S, Huber A, Zöllner N (1988) Four DNA polymorphisms in the LDL-receptor gene and its use in diagnosis of familial hypercholesterolemia. (Submitted)
33. Antonarkis SE, Kazazian HH, Orkin SH (1985) DNA polymorphism and molecular pathology of the human globin gene clusters. Hum Genet 69:1–14
34. Conner BJ, Reyes AA, Morin C, Itakura K, Teplitz RL, Wallace RB Detection of sickle cell βS globin allele by hybridization with synthetic oligonucleotides. Proc Natl Acad Sci USA 80:278–282
35. Funke H, Rust S, Assmann G (1986) Detection of apolipopreotein E variants by an oligonucleotide melting procedure. Clin Chem 32,7:1285–1289
36. Saiki RK, Scharf S, Faloona F, Mullis KB, Horn GT, Erlich HA, Arnheim N (1985) Enzymatic amplification of β-globin genomic sequences and restriction site analysis for diagnosis of sickle cell anemia. Science 230:1350–1354
37. Brink PA, Steyn LT, Coetzee GA, Van der Westhuyzen DR (1987) Familial hypercholesterolaemia in South African Afrikaners: PvuII and StuI DNA polymorphisms in the LDL-receptor gene consistent with a predominant founder gene effect. Hum Genet (In press)

Molekularbiologie und klinische Forschung: das Beispiel des LDL-Rezeptors und der Familiären Hypercholesterinämie

T. C. Südhof

Die Entdeckung des LDL-Rezeptors ist einem echten Experiment der klinischen Forschung zu verdanken: Brown und Goldstein wollten eigentlich die Regulation des Enzyms HMG-Coenzym A-Reduktase untersuchen und führten ihre Experimente mit der Hypothese durch, daß der primäre Defekt bei der familiären Hypercholesterinämie (FH) in einer gestörten Regulation der Cholesterinbiosynthese läge (Brown et al. 1973; Brown u. Goldstein 1974; Brown et al. 1981). Die Experimente dagegen wiesen in die entgegengesetzte Richtung: sie zeigten, daß normale Zellen einen spezifischen Rezeptor für cholesterinreiche Lipoproteine haben, und daß dieser Rezeptor in Zellen von Patienten mit FH fehlte. In diesen Zellen wurde die Cholesterinbiosynthese also nicht aufgrund eines Regulationsdefektes aktiviert, sondern sie wurde aktiviert, gerade weil die Regulation funktionierte: die Zellen erhielten kein Cholesterin aus dem Medium, also mußten sie es selber synthetisieren (Brown et al. 1974).

Das Thema dieses Artikels soll der Einfluß der Molekularbiologie auf die biomedizinische Forschung sein, dargestellt am Beispiel der familiären Hypercholesterinämie und des LDL-Rezeptors. Es ist wichtig zu bemerken, daß auf diesem Gebiet die wesentlichen biologischen und medizinischen Konzepte bereits formuliert worden waren, als die Molekularbiologie als neue revolutionäre Technologie entwickelt wurde. Auf die Entdeckung der Existenz des LDL-Rezeptors als spezifischen und ubiquitären Oberflächenrezeptor für Lipoproteine folgte zunächst seine detaillierte biochemische Charakterisierung, die in der Erzeugung spezifischer monoklonaler Antikörper gegen den Rezeptor (Beisiegel et al. 1981 a, b) und in der biochemischen Reinigung des Rezeptors gipfelte (Schneider et al. 1982). Die biochemische Charakterisierung erlaubte gleichzeitig eine Untersuchung der Pathologie des LDL-Rezeptors bei Patienten mit FH, die zu der klaren Demonstration führte, daß bei all diesen Patienten der LDL-Rezeptor funktionell fehlt, aber daß die biochemische Basis dieses Defektes heterogen ist (Tolleshaug et al. 1982; Goldstein et al. 1983). Parallel mit der biochemischen Erforschung des LDL-Rezeptors ging seine zellbiologische Charakterisierung. Mit der Verfügbarkeit spezifischer Reagenzien gegen den Rezeptor konnte sein intrazellulärer Weg untersucht werden. Diese Studien führten zur Entdeckung der rezeptormediierten Endozytose und der entscheidenden Rolle von Clathrin im intrazellulären Membranverkehr (Anderson et al. 1976, 1977, 1982). Die rezeptormediierte Endozytose ist inzwischen zu einem der Paradigmen der Zellbiologie geworden mit der Erkenntnis, daß die Mehrheit der Rezeptoren und viele andere Membranproteine diesen Mechanismus zur Endozytose benutzen (Brown et al. 1983).

Die grundsätzlichen Konzepte, die dieser Forschung entstammten, nämlich die Existenz von spezifischen Oberflächenrezeptoren für Lipoproteine, die Bedeutung der Blutkonzentration von Lipoproteinen für die Entstehung der Atherosklerose, und die rezeptormediierte Endozytose als generellem zellbiologischen Phänomen, waren also bereits vorhanden, als die molekularbiologische Technik möglich wurde. Trotzdem hatte diese einen tiefen Einfluß auf die Weiterentwicklung dieser Konzepte und auf die Formulierung neuer biologischer Hypothesen. Als erstes erlaubte es die Klonierung des LDL-Rezeptors, seine Primärsequenz zu untersuchen (Yamamoto et al. 1984). Darauf erfolgte die Klonierung des humanen Gens für den LDL-Rezeptor (Südhof et al. 1985a, b), dessen Struktur wichtige Schlußfolgerungen zuließ: Die Struktur des Gens ermöglichte eine Unterteilung des LDL-Rezeptors in funktionell distinkte Domäne, welche für die Strukturfunktionsuntersuchungen am LDL-Rezeptor wegweisend waren. Diese Widerspiegelung der Domänstruktur des LDL-Rezeptorproteins in der Genstruktur erwies sich als das bisher beste Beispiel für "exon shuffling" in der Evolution, einer Hypothese, die die Existenz von Exonen und Intronen im Genom erklärt, indem sie die Exone gewissermaßen als die Steine im Mosaik vieler Proteine ansieht (Gilbert 1978, 1985). Weiterhin ermöglichte die Charakterisierung des LDL-Rezeptorgens die molekulare Untersuchung der Mutationen bei Patienten mit FH, die seither in großem Maßstab vorangetrieben worden ist und zu vielen Einsichten in die Biologie des LDL-Rezeptors und in die molekularen Grundlagen genetischer Krankheiten geführt hat (Lehrman et al. 1985; Russell et al. 1986). Es konnte gezeigt werden, daß die Mutationen im LDL-Rezeptorlokus genetisch heterogen sind, d. h. das Ergebnis vieler unabhängiger Mutationsereignisse waren, obwohl bestimmte Mutationen sich in einzelnen isolierten Populationen anreichern konnten, wie z. B. in den "French Canadians" (Hobbs et al. 1987). Die molekulare Charakterisierung vieler Mutationen deutete auf eine Beteiligung repetitiver Sequenzen bei der Genese von Deletionen hin, einem pathogenen Mechanismus, der von allgemeiner Bedeutung für die Entstehung genetischer Erkrankungen sein konnte (Lehrman et al. 1985). Diese Ergebnisse zusammen gaben außerdem Anlaß zu systematischen Transfektionsexperimenten, in denen normale und mutierte LDL-Rezeptoren in LDL-rezeptornegativen Zellen exprimiert wurden (Davis et al. 1987; van Driel et al. 1988). Die Transfektionsexperimente ergaben eine funktionelle Abgrenzung der Domäne des LDL-Rezeptors und erlaubten u. a. die Identifizierung der Ligandenbindungsstelle des Rezeptors.

Eines der interessantesten Phänomene des Cholesterinstoffwechsels ist die Regulation der Expression der beteiligten Proteine durch Cholesterin. So wird z. B. die Expression des LDL-Rezeptors im Organismus genau durch die Menge des vorhandenen Cholesterins kontrolliert und kann durch medikamentöse Hemmung der Cholesterinbiosynthese induziert werden (Ma et al. 1986). Die Regulation des LDL-Rezeptors ist von nicht zu unterschätzender potentieller klinischer Bedeutung, da eine ständige Suppression der LDL-Rezeptorexpression funktionell das gleiche ist wie die familiäre Hypercholesterinämie. Wenn also z. B. 10% der Bevölkerung eine verhältnismäßig stärkere Suppression des LDL-Rezeptors nach Cholesterinbelastung zeigen, so ist wahrscheinlich, daß diesen Bevölkerungsteil eine besonders hohe Inzidenz von Atherosklerose treffen wird.

Die Mechanismen der cholesterinbedingten Regulation der LDL-Rezeptorexpression wurden mit molekularbiologischen Methoden untersucht (Südhof et al. 1987 a, b). Es konnte gezeigt werden, daß diese Regulation allein auf der Ebene der Gentransskription erfolgt. Die Promotersequenzen, die für diese Regulation verantwortlich sind, konnten lokalisiert werden. Eine kurze, 15 Basenpaare lange Sequenz im Promoter des 50 kb langen LDL-Rezeptorgens stellt die Erkennungssequenz für die Regulation dar. Ähnliche Sequenzen mit gleichen Eigenschaften konnten auch in anderen cholesterinregulierten Genen gefunden werden. Wenn diese Sequenzen in ein anderes, nicht cholesterinabhängiges Gen transferiert werden, kann dieses Gen durch Cholesterin supprimierbar gemacht werden (Südhof et al. 1987 b). Diese Studien erlauben jetzt die Untersuchung der Proteinfaktoren, die an dieser Regulation beteiligt sind (Dawson et al. 1987). Eine weitere wichtige, aktiv beforschte Frage ist, ob Individuen verschieden stark regulativ auf Cholesterin reagieren und ob diese Unterschiede genetisch lokalisiert werden können.

Was für Schlußfolgerungen können aus diesen Beispielen gezogen werden über den Einfluß der Molekularbiologie auf die Entwicklung der Forschung? Vielleicht gilt generell, was für dieses Gebiet im besonderen gezeigt wurde: Grundlegende biologische Konzepte wurden durch die Technik per se nicht aufgedeckt, jedoch konnten die bestehenden Konzepte schneller verifiziert oder falsifiziert und mit Mechanismen versehen werden. Wegweisend in neue Richtungen wurde die molekularbiologische Technik dann, wenn methodisch nicht anders angehbare Fragen thematisiert wurden. Dann allerdings führten überraschende Ergebnisse zu neuen Perspektiven.

References

1. Anderson RGW, Goldstein JL, Brown MS (1976) Localization of low density lipoprotein receptors on plasma membrane of normal human fibroblasts and their absence in cells from a familial hypercholesterolemia homozygote. Proc Natl Acad Sci USA 73:2434–2438
2. Anderson RGW, Brown MS, Goldstein JL (1977) Role of the coated endocytic vesicle in the uptake of receptor-bound low density lipoprotein in human fibroblasts. Cell 10:351–364
3. Anderson RGW, Brown MS, Beisiegel U, Goldstein JL (1982) Surface distribution and recycling of the low density lipoprotein receptor as visualized with antireceptor antibodies. J Cell Biol 93:523–531
4. Beisiegel U, Kita T, Anderson RGW, Schneider WJ, Brown MS, Goldstein JL (1981 a) Immunologic cross-reactivity of the low density lipoprotein receptor from bovine adrenal cortex, human fibroblasts, canine liver and adrenal gland, and rat liver. J Biol Chem 256:4071–4078
5. Beisiegel U, Schneider WJ, Goldstein JL, Anderson RGW, Brown MS (1981 b) Monoclonal antibodies to the low density lipoprotein receptor as probes for study of receptor-mediated endocytosis and the genetics of familial hypercholesterolemia (1981 b). J Biol Chem 256:11923–11931
6. Brown MS, Dana SE, Goldstein JL (1973) Regulation of 3-hydroxy-3-methylglutaryl coenzyme A reductase activity in human fibroblasts by lipoproteins. Proc Natl Acad Sci USA 70:2162–2166
7. Brown MS, Goldstein JL (1974) Familial hypercholesterolemia: defective binding of lipoproteins to cultured fibroblasts associated with impaired regulation of 3-hydroxy-3-methylglutaryl coenzyme A reductase activity. Proc Natl Acad Sci USA 71:788–792
8. Brown MS, Dana SE, Goldstein JL (1974) Regulation of 3-hydroxy-3-methylglutaryl coenzyme A reductase activity in cultured human fibroblasts. J Biol Chem 249:789–796
9. Brown MS, Kovanen PT, Goldstein JL (1981) Regulation of plasma cholesterol by lipoprotein receptors. Science 212:628–635
10. Brown MS, Anderson RGW, Goldstein JL (1983) Recycling receptors: the round-trip itinerary of migrant membrane proteins. Cell 32:663–667
11. Davis CG, Goldstein JL, Südhof TC, Anderson RGW, Russell DW, Brown MS (1987) Acid-dependent ligand dissociation and recycling of LDL receptor mediated by growth factor homology region. Nature 326:760–764
12. Dawson PA, Hofmann SL, van der Westhuyzen DR, Südhof TC, Brown MS, Goldstein JL (1988) Sterol-dependent repression of low density lipoprotein receptor promoter mediated by 16-base pair sequence adjacent to binding site for transcription factor spl. J Biol Chem 263:3372–3379
13. Gilbert W (1978) Why genes in pieces? Nature (London) 271:501
14. Gilbert W (1985) Genes-in-pieces revisited. Science 228:823
15. Goldstein JL, Kita T, Brown MS (1983) Defective lipoprotein receptors and atherosclerosis. New England J of Medicine 309:288–296
16. Hobbs HH, Brown MS, Russell DW, Davignon J, Goldstein JL (1987) Deletion in the gene for the low-density-lipoprotein receptor in a majority of french canadians with familial hypercholesterolemia. New England J of Medicine 317:734–737
17. Lehrman MA, Schneider WJ, Südhof TC, Brown MS, Goldstein JL, Russel DW (1985) LDL receptor mutation: alu-alu recombination deletes exons encoding transmembrane and cytoplasmic domains. Science 227:140–146
18. Ma PT, Gil G, Südhof TC, Bilheimer DW, Goldstein JL, Brown MS (1986) Mevinolin, an inhibitor of cholesterol synthesis, induces mRNA for low density lipoprotein receptor in livers of hamsters and rabbits. Proc Natl Acad Sci USA 83:8370–8374

19. Russell DW, Lehrman MA, Südhof TC, Yamamoto T, Davis CG, Hobbs HH, Brown MS, Goldstein JL (1986) The LDL receptor in familial hypercholesterolemia: use of human mutations to dissect a membrane protein. Cold Spring Harbor Symp Quant Biol LI:811–819
20. Schneider WJ, Beisiegel U, Goldstein JL, Brown MS (1982) Purification of the low density lipoprotein receptor, and acidic glycoprotein of 164,000 molecular weight. J Biol Chem 257:2664–2673
21. Südhof TC, Goldstein JL, Brown MS, Russell DW (1985a) The LDL receptor gene: a mosaic of exons shared with different proteins. Science 228:815–822
22. Südhof TC, Russell DW, Brown MS, Goldstein JL, Sanchez-Pescador R, Bell GT (1985b) Cassett of eight exons shared by genes for LDL receptor and EGF precursor. Science 228:893–895
23. Südhof TC, Russell DW, Brown MS, Goldstein JL (1987a) 42-bp element from LDL receptor gene confers end-product repression by sterols when inserted into viral TK promoter (1987a). Cell 48:1061–1069
24. Südhof TC, Westhuyzen DVD, Goldstein J, Michael B, Russell D (1987b) Three direct repeats and a TATA-like sequence are required for regulated expression of the human LDL receptor gene. J Biol Chem 262:10773–10779
25. Tolleshaug H, Goldstein JL, Schneider WJ, Brown MS (1982) Posttranslational processing of the LDL receptor and its genetic distruption in familial hypercholesterolemia. Cell 30:715–724
26. van Driel IR, Goldstein JL, Südhof TC, Brown MS (1987) First cystein-rich repeat in ligand-binding domain of low density lipoprotein receptor bind Ca^{2+} and monoclonal antibodies, but not lipoproteins. J Biol Chem 262:17443–17449
27. Yamamoto T, David CG, Brown MS, Schneider WJ, Casey ML, Goldstein JL, Russell DW (1984) The human LDL receptor: a cysteine-rich protein with multiple alu sequences in its mRNA. Cell 39:27–38

Genetics, Drugs and Low Density Lipoprotein Metabolism

C. J. Packard and J. Shepherd

Introduction

Low density lipoprotein (LDL), the major carrier of cholesterol in human plasma, occupies a central role in sterol metabolism and in the pathological processes that lead to atherosclerosis and ischaemic heart disease. It has been amply demonstrated that plasma concentrations of LDL cholesterol above 180 mg/dl (4.5 mmol/l) are associated with accelerated atherogenesis. Conversely, reducing LDL cholesterol levels in high risk individ-

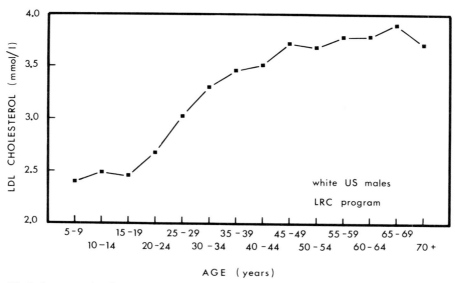

Fig. 1. Increment in plasma LDL with age. The data, taken from the Lipid Research Clinics Prevalence Study, describe the rise in plasma LDL in white US males

uals halts the progression of lesions or even promotes their regression. Despite the fact that LDL has been intensively studied for 30 years, our knowledge of how its circulating level is controlled is still only rudimentary. For example, the concentration of LDL in the population increases by up to 40% in the third and fourth decades of life (Fig. 1). The reason for this dramatic change with its consequences for coronary disease is unknown, although we have evidence that it is not an obligatory feature of aging since not all national groups show the effect. In the discussion that follows we will

1. Summarise current understanding of the control of LDL metabolism and the influence of therapeutic agents
2. Explore the progress that has been made in elucidating the structural and metabolic heterogeneity in LDL
3. Review recent results that shed light on the influence of genetic variation on LDL metabolism

Regulation of Plasma LDL Concentration

LDL concentrations vary widely in the population, the level of cholesterol in this fraction commonly fluctuating between 2.0 and 6.0 mmol/l. This plasma constituent is therefore poorly regulated compared with others, such as sodium or even albumin. If we assume that "healthy" levels are those seen in young adults, then a plasma concentration of 2.0–3.0 mmol/l of LDL cholesterol would be considered ideal. Metabolic studies [1] indicate that as individuals age a combination of increased synthesis and reduced catabolism lead to higher LDL levels. In fact, in a survey of normal and hyperlipaemic subjects we found that synthesis and catabolism were of equal importance in determining the level of circulating LDL [2]. Furthermore, the major organ involved in both the production and breakdown of LDL is the liver, and so the hepatocyte is the key cell in the regulation of plasma LDL cholesterol concentrations.

Regulation of Catabolism

The discovery of a specific membrane protein that binds LDL and promotes its internalisation and degradation was pivotal to our current appreciation of LDL catabolism [3]. This "LDL receptor" is distributed widely throughout the cytoplasmic membranes of most tissues. When it appears on a cell's surface, it facilitates endocytosis of LDL particles by binding them to specialised "coated pit" areas of membrane that undergo rapid endocytosis. The LDL is passed in an endocytic vesicle to the perinuclear region where, following fusion of the vesicle with lysosomes, the transported LDL is broken down to release cholesterol for the cell's need. Once the cell has acquired sufficient sterol, receptor synthesis is downregulated and the whole pathway is shut down. Receptor-mediated LDL catabolism is quantitatively important in lipoprotein catabolism in man, accounting for 50%–70% of total LDL clearance in normal subjects. The remainder is degraded by poorly understood receptor-independent mechanisms that are thought to be integral to cells of the monocyte-macrophage system which, while operating as general scavengers in the body, are also believed to play a key role in atherogenesis.

Subjects with particularly elevated levels of LDL such as those with familial hypercholesterolaemia (FH) or hypothyroidism have inherited or acquired receptor deficiency that leads to an imbalance in LDL catabolism [4, 5]. For example, in heterozygous FH where half the normal complement of receptors is missing or non-functional, clearance via the receptor route accounts for only 16% of total catabolism. The burden of removal therefore falls on the receptor independent mechanisms, a fact that helps explain the very high incidence of premature and severe ischaemic heart disease in these individuals. Likewise, hypothyroid patients exhibit defective catabolism of LDL which is normalised when they are given replacement thyroxine.

Knowledge of the operation of the LDL receptor pathway also provides an explanation for the mechanism of action of commonly used lipid lowering agents. Cholestyramine, the agent used in the successful Lipid Research Clinics (LRC) Coronary Primary Prevention Trial, is not absorbed by the body, but acts in the intestinal lumen by binding bile acids and promoting their faecal loss. This has a secondary effect on hepatic sterol metabolism since bile acids are derived by oxidation of cholesterol in the liver. The organ compensates for the several-fold increase in faecal excretion by increasing activity of the pacemaker enzyme cholesterol 7 α hydroxylase, drawing substrate sterol in part from increased cholesterologenesis and partly by accelerating the synthesis of LDL receptors [6]. These proteins promote the import of LDL cholesterol into the liver and so plasma levels fall, usually by 20%–25% at maximal drug dosage. This increase in hepatic receptor-mediated LDL catabolism in the LRC trial was associated with a reduced incidence of coronary events and so appears to be a beneficial way of correcting the imbalance between receptor-dependent and receptor-independent mechanisms in hypercholesterolaemic subjects.

A second class of drugs is believed to have a similar mechanism of action. The 3-hydroxy-3-methylglutaryl coenzyme A (HMG CoA) reductase inhibitors act upon the pacemaker enzyme in cholesterol synthesis to reduce de novo production of sterol, especially in the liver. The organ, in the face of continued demand for cholesterol, again draws LDL from the circulation via a stimulated receptor pathway [7]. Commonly, drugs in this category (like lovastatin, pravastatin), when given as primary agents to hypercholesterolaemic subjects, induce decrements of 30% in LDL cholesterol; and as might be expected on theoretical grounds, they are also particularly potent supplements to bile acid sequestrant resins. In this context they blunt the increase in hepatic cholesterol synthesis which commonly follows resin therapy. So, the liver becomes even more dependent on plasma LDL as its sterol source and reductions of 50% in circulating LDL cholesterol can be observed in subjects given combinations of resin and reductase inhibitor therapy.

Regulation of Synthesis

In contrast to the advances in our understanding of LDL catabolism during the last 10 years, appreciation of how synthesis is regulated and of the mechanisms by which precursor very low density lipoprotein (VLDL) is converted to LDL is still largely rudimentary. It is known that in normal subjects all LDL apolipoprotein B (apo B) is derived from VLDL catabolism, but this relationship is disrupted in hyperlipidaemia. Subjects with hypertriglyceridaemia secrete more apo B into VLDL than ultimately appears in LDL. A high proportion of this VLDL is therefore removed directly without becoming LDL [8]. Most homozygous familial hypercholesterolaemic patients, on the other hand, produce LDL directly by VLDL-independent pathways [9]. We have investigated the VLDL-LDL conversion process in detail [10] and have concluded that the metabolic cascade is more complex than was first envisioned, even in normal individuals. When VLDL is fractionated into large (S_f 60–400) and small (S_f 20–60)

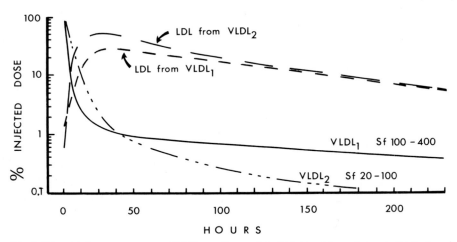

Fig. 2. Plasma clearance curves of LDL apolipoprotein B derived from large (*solid line*) and small (*dotted line*) VLDL (S_f 100–400 and 60–100 respectively)

components and the fates of both are traced in vivo, it can be seen that the extent and rate of conversion of the smaller species to LDL is greater than that of the larger. About 10%–20% of apo B in large VLDL finds its way to LDL, whereas 50%–60% of small VLDL becomes LDL. Furthermore, the metabolic behaviour of LDL apo B derived from large VLDL differs significantly from that of LDL derived from small VLDL. The former appears more slowly and is catabolised less efficiently than the latter (Fig. 2). This indicates that LDL is metabolically heterogeneous (as discussed below).

Metabolic Heterogeneity of LDL

LDL isolated from plasma by density gradient ultracentrifugation appears to be composed of one major species and a few minor fractions. When the kinetics of this major fraction are determined in vivo, it appears that it, too, is heterogeneous [11]. Estimation of the daily catabolic rate of trace-labelled LDL by the urine/plasma (U/P) radioactivity ratio (Fig. 3) shows that this decreases during the period of the turnover. Ideally, if a tracer were homogeneous, this daily U/P ratio would remain unchanged, indicating that a constant fraction of the plasma pool was being catabolised each day. A decrease in U/P of up to 50% as seen in some of our studies indicates that the tracer is composed of at least two fractions, one of which is degraded 2–3 times faster than the other (Fig. 3). This phenomenon has been shown to occur in most normal and hyperlipaemic subjects [11]. The

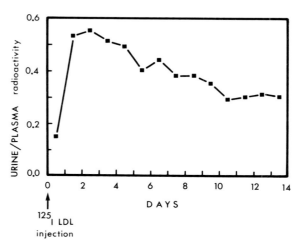

Fig. 3. Changes in daily U/P radioactivity ratios in a normal subject following an intravenous injection of ^{125}I LDL

nature of the two LDL species is not clear. However, more recent work of ours and others [12] has indicated that the rapidly catabolised component is more efficiently removed by receptor mechanisms than the slower one. This conclusion is based on the finding that when a subject is injected simultaneously with ^{125}I-native LDL and ^{131}I-chemically modified LDL which does not bind to receptors, the U/P ratio for the former tracer decreases from a high value over the 14-day study, but the latter is relatively flat. Hence, during the initial phase the receptor contribution (native LDL clearance minus receptor-blocked LDL clearance) is about 75% of the total catabolic activity, while the receptors account for about half of total catabolism during the terminal phase. Further evidence for the role of receptors in this catabolic difference comes from the observation that individuals with heterozygous familial hypercholesterolaemia, who have a partial deficiency of LDL receptors, exhibit a flat U/P curve for both native and chemically modified LDL.

The metabolic heterogeneity in LDL must reflect some basic structural diversity. As mentioned above, it is possible to isolate by standard rate zonal ultracentrifugation methods a major LDL species that appears to be homogeneous in terms of size and yet displays gross metabolic heterogeneity. The discrepancy indicates that large differences in lipid content are probably not the basis of the differences between LDL subspecies since this would give rise to an altered size and density. A more likely explanation lies in finding that some LDL fractions contain more apo E per particle than others. Lee and Alaupovic [13] have reported that approximately one-third of LDL contains 0.74 mol of apolipoprotein E (apo E) per mol apo B, while the remainder contains 0.08 mol of apo E per mol apo B. It is known that apo E has about a 10-times higher affinity for the LDL receptor than apo B and so its presence, even in small amounts, on a lipoprotein particle may give rise to a much-enhanced rate of clearance. This hypothesis could be readily tested by determining the rate of fall in the U/P ratio of subjects who have the abnormal E_2 variant which fails to bind efficiently to receptors.

LDL Metabolism in Hypertriglyceridaemia

Abnormalities in LDL metabolism occur not only in hypercholesterolaemic subjects who have supranormal amounts of the lipoprotein, but also in individuals with high triglyceride levels (> 3.0 mmol/l) who show subnormal LDL cholesterol levels due to a combination of structural and metabolic changes. Lipoproteins are able to interchange their core triglyceride and cholesteryl esters via the agency of a "neutral lipid transfer protein". If triglyceride is present in large amounts in chylomicrons and VLDL, then this will be transferred to the denser lipoproteins LDL and

HDL. Conversely, cholesteryl ester will move from the latter into chylomicrons and VLDL. Recent studies [14] have shown that triglyceride enrichment of LDL and HDL has important consequences for their structure and metabolism. These lipoproteins become increasingly better substrates for the enzyme hepatic lipase which hydrolyses the triglyceride, reducing the core volume and hence the size of the lipoprotein particle. Thus, subjects with high triglyceride levels have smaller, denser LDL than normal. Recent studies have shown that hypertriglyceridaemic LDL manifests defective binding to the LDL receptor on cultured fibroblasts [15]. This is possibly due to an altered conformation of apoprotein B on the lipoprotein particle's surface. Correction of the hyperlipaemia with lipid-lowering drugs such as bezafibrate results in the plasma accummulation of LDL that is more normal in composition and that exhibits a restored binding capacity. The structure of apo B on its surface, as detected by monoclonal antibodies, is altered during treatment [16]. The altered functionality of LDL in hypertriglyceridaemic subjects should have important consequences for its metabolism, and indeed, marked abnormalities are evident in kinetic studies.

Originally the low LDL in subjects with high triglyceride levels was thought to be due to the failure of VLDL to be converted to LDL. Studies of apo B turnover, however, indicated that the synthetic rate of LDL was normal in these individuals, but the catabolic rate was increased often substantially above the value seen in controls (e.g. 0.45 pools/day in hypertriglyceridaemic subjects vs 0.35 pools/day in controls). The reason for this hypercatabolism was examined in our laboratory in a study in which hypertriglyceridaemic patients were injected with native and chemically modified LDL in order to determine the contributions of the receptor-dependent and receptor-independent pathways [17]. We found that the amount of apo LDL cleared by the receptor route in these subjects was subnormal (3.9 mg/kg per day versus 6.0 mg/kg per day in normal subjects) while the burden on the receptor independent route was increased (9.0 mg/kg per day versus 7.7 mg/kg per day in normal individuals). The cause for this imbalance may be the altered structure of the LDL, which may have a lower affinity for receptors due to changes in apo B conformation, while the decrease in size may promote receptor-independent clearance. Certainly, the fraction of the plasma pool removed by non-receptor catabolism in hypertriglyceridaemic patients was twice that in the normal individuals. Treatment with fibrates (fenofibrate in this instance) corrected these metabolic abnormalities and at the same time reversed the structural perturbations described above. Both the amount cleared by the receptor route and that degraded by non-receptor mechanisms fell into the normal range.

Genetics and Structural Heterogeneity in LDL

Considerable attention is being focussed at present on the structural heterogeneity within LDL. Classical techniques such as density gradient centrifugation generally fail to resolve LDL into major subspecies. Newer methodologies, particularly non-denaturing gradient gel electrophoresis, reveal that LDL is, in most individuals, composed of a small number of closely related species. How these arise or which mechanisms are responsible for their interconversion and catabolism is unknown. Preliminary studies of the LDL subclass pattern in families suggest that it is influenced strongly by inheritance. Austin and Krauss [18] have shown that variation at a single gene locus contributes to the pattern. The LDL phenotype is also influenced by age, sex and plasma triglyceride level. How these interrelate is still to be established.

Genetics and LDL Metabolism

Plasma cholesterol and particularly LDL cholesterol levels are influenced by both genetic and environmental factors. Inheritance studies of dizygotic and monozygotic twins have indicated that about half of the variation in cholesterol is genetic in origin, while the remainder is attributable to environment, and in particular to diet. Two major genetic polymorphisms have been detected which may play a role in regulation LDL [19]. The first is associated with apo E, a minor apoprotein, which is nevertheless important in promoting receptor-mediated uptake of lipoproteins (see above). Individuals inheriting two copies of the E_2 mutation, in which the arginine at amino acid 145 is replaced with a cysteine, have lower levels of LDL cholesterol than those who express the common E_3 variety. Conversely, those homozygous for the E_4 allelic variant (where cys_{158} is replaced with an arginine) have higher LDL cholesterol levels than E_3 homozygotes. These mutations in apo E have been estimated to contribute about 15% to the variability in LDL in the general population. The reason for the effect is not known. A number of possibilities have been proposed. Davignon et al. [19] have suggested that the possession of the E_4 protein leads to improved uptake of cholesterol-carrying chylomicron remnants by the liver. This in turn increases hepatic cholesterol levels, reduces cellular receptor activity and diminishes the catabolism of LDL by the organ. As a result, LDL levels increase above those seen in E_3 homozygotes. The reciprocal mechanism holds in subjects with the E_2 protein, in whom defective chylomicron clearance produces enhanced LDL hepatic uptake. An alternative explanation for the lower LDL levels in apo E_2 homozygotes has arisen from metabolic studies of the VLDL-LDL conversion in our laboratory. Individuals with the E_2/E_2 phenotype convert less VLDL to LDL than E_3/E_3 subjects and so decreased synthesis is the primary cause of lower plasma levels rather than accelerated clearance (Demant, Packard and Shepherd, unpublished results).

The second genetic variation that has been reported to alter plasma LDL levels is linked to the gene for apo B. Here a restriction fragment length polymorphism has been detected using the enzyme XbaI and the genomic probe pABC. Individuals with the cutting site for the enzyme (designated X_2/X_2) have higher levels of LDL than those without the restriction site (genotype X_1X_1) [20]. The polymorphism is not present in a coding region of the gene, but since the mutation affects the plasma concentration of LDL, the site must be closely linked to another region with important properties for controlling either the production or degradation of the lipoprotein. In fact, the restriction site is adjacent to the putative ligand binding region of apo B, and variations in receptor binding may be an explanation for the altered plasma levels. This hypothesis receives support from recent studies of LDL metabolism in these genotypic variants. Individuals expressing the X_2X_2 genotype have a slower catabolic rate of apo LDL than those with X_1X_1 [21, 22]. Furthermore, this reduction in catabolic efficiency is attributable to decreased degradation by the receptor pathway, indicating that impaired ligand-receptor interaction may be responsible for the differences in LDL levels. Clearly, if the apo E and apo B variations are considered together, they must account for a substantial proportion of the genetic influence which bears on LDL levels. Other polymorphisms in apo B will undoubtedly be uncovered and contribute further to our understanding of the regulation of plasma LDL metabolism.

LDL and Atherosclerosis

High LDL levels are associated with accelerated atherosclerosis, but the mechanisms underlying this association are largely unknown and only recently [23] has some progress been made in understanding how elevations in LDL may promote cholesterol deposition in atheromatous lesions. Experiments in tissue culture have shown that if LDL is incubated with endothelial or smooth muscle cells, it undergoes a series of changes that result in the production of a particle with increased electronegative charge and a lower cholesterol/protein ratio. The modified LDL, unlike the native lipoprotein, causes intracellular cholesterol ester deposition when presented to macrophages. Since these cells are found in the subendothelial space and are believed to be intimately involved in the atherogenic process, it is possible to envisage that the passage of LDL through the endothelial layer may result in the production of electronegative LDL which is taken up by wandering subendothelial monocytes that form the focus of a nascent lesion. Further studies [24] have indicated that oxidation of the LDL may be the cause of these changes in the particle's structure and behaviour. It is known that superoxide radicals are generated by endothelial and smooth muscle cells, and free radical scavengers in vitro can inhibit the above changes in LDL. Furthermore, the drug probucol, which acts as a free radical scavenger, can apparently inhibit LDL uptake by atherosclerotic lesions in vivo. Clearly, further examination of this aspect of LDL metabolism may be fruitful in revealing novel areas of potential therapeutic importance.

References

1. Miller NE (1984) Lancet I:263–266
2. Packard CJ, Shepherd J (1983) Atheroscler Revs 11:29–63
3. Goldstein JL, Brown MS (1977) Annu Rev Biochem 46:897–930
4. Shepherd J, Bicker S, Lorimer AR, Packard CJ (1979) J Lipid Res 20:999–1006
5. Thompson GR, Soutar AK, Spengel FA, Jadhav A, Gavigan SJP, Myant NB (1981) Proc Natl Acad Sci USA 78:2591–2595
6. Packard CJ, Shepherd J (1982) J Lipid Res 23:1081–1098
7. Bilheimer DW, Grundy SM, Brown MS, Goldstein JL (1983) Proc Natl Acad Sci USA 80:4124–4128
8. Reardon MF, Fidge NH, Nestel PJ (1978) J Clin Invest 61:850–860
9. Soutar AK, Myant NB, Thompson GR (1977) Atherosclerosis 28:247–256
10. Packard CJ, Munro A, Lorimer AR, Gotto AM, Shepherd J (1984) J Clin Invest 74:2178–2192
11. Foster DM, Chait A, Albers JJ, Failor A, Harris C, Brunzell JD (1986) Metabolism 35:658–696
12. Beltz WF, Young SG, Witztum JL (1987) In: Proceedings of the workshop on lipoprotein heterogeneity. NIH Public No 87–2646, pp 215–236
13. Lee DM, Alaupovic P (1986) Biochim Biophys Acta 879:126–133
14. Eisenberg S, Gavish D, Oschry Y, Fainaru M, Deckelbaum RJ (1984) J Clin Invest 74:470–482
15. Kleinman Y, Oschry Y, Eisenberg S (1987) Eur J Clin Invest 17:538–543
16. Kleinman Y, Schonfeld G, Gavish D, Oschry Y, Eisenberg S (1987) J Lipid Res 28:540–548
17. Shepherd J, Caslake MJ, Lorimer AR, Vallance BD, Packard CJ (1985) Arteriosclerosis 5:162–168
18. Austin MA, Krauss RM (1986) Lancet II:592–595
19. Davignon J, Gregg RE, Sing CF (1988) Arteriosclerosis 8:1–21
20. Law A, Powell LM, Brunt H, Knott TJ, Altman DG, Rajput I, Wallis SC, Pease RJ, Priestley LM, Scott J, Miller GJ, Miller NE (1986) Lancet:1301–1304
21. Demant T, Houlston RS, Caslake MJ, Series JJ, Shepherd J, Packard CJ, Humphries SE (1988) J Clin Invest (in pres)
22. Houlston RS, Turner PR, Revill J, Lewis B, Humphries SE (1988) Atherosclerosis 71:81–85
23. Henriksen T, Mahoney EM, Steinberg D (1983) Arteriosclerosis 3:149–159
24. Morel DW, DiCorleto PE, Chisholm GM (1984) Arteriosclerosis 4:357–364

Visualization and Characterization
of the Scavenger Receptor in Human Liver *, **

K. HARDERS-SPENGEL

Introduction

The atherogenic potential of elevated serum cholesterol levels is nowadays well established. Persons with hypercholesterolemia, usually in the form of elevated low density lipoprotein (LDL) levels, run a higher risk of developing atherosclerotic lesions than normocholesterolemic subjects of the same age. Patients with the autosomal dominant trait of familial hypercholesterolemia (FH) exhibit dramatically elevated LDL levels and are prone to massive and premature atherosclerosis. Untreated patients homozygous for FH frequently die of myocardial infarction in early childhood (Table 1).

The classic studies of Goldstein and Brown and many others have shown the LDL receptor which mediates specific cellular uptake and degradation of LDL to be defective or absent in patients with FH (Brown and

Table 1. Patients homozygous for FH. Data compiled from the literature

Number of patients	196
Mean serum cholesterol level (mg/dl)	742
Onset of angina pectoris (mean age, years)	11
First myocardial infarction (mean age, years)	14
Death from myocardial infarction (mean age, years)	15

* This paper is dedicated to Professor Dr. Nepomuk Zöllner on the occasion of his 65th birthday. I am grateful for his continued encouragement and support over the last 10 years.

** The author is a recipient of grants from the Deutsche Forschungsgemeinschaft and from the Wissenschaftliches Herausgeberkollegium der Münchener Medizinischen Wochenschrift e. V.

29

Goldstein 1986). This defect is also demonstrable in the liver, in normal subjects otherwise the principal site of LDL catabolism (Spengel et al. 1981; Thompson et al. 1981). Specific LDL binding to liver membranes from patients heterozygous for FH is reduced by 56% as compared with the specific binding of LDL to liver membranes from normocholesterolemic non-FH subjects (Harders-Spengel et al. 1982). Though defective or absent LDL receptors lead to hypercholesterolemia via reduced catabolism, the LDL receptor pathway cannot be directly responsible for atherogenesis since (1) especially patients homozygous for FH, who predominantly are the victims of massive and premature atherosclerosis, completely or nearly completely lack functioning LDL receptors, and (2) LDL receptor-dependent endocytosis of LDL is regulated by a sensitive and intricate feedback inhibition that prevents cholesterol overload of the cell. However, a positive relationship between plasma cholesterol levels and cholesterol content of the artery wall has been shown by many authors (Friedman et al. 1962; Niehaus et al. 1977; Smith and Slater 1972; Smith 1974, 1977; Smith and Staples 1980; Stender and Zilversmit 1982).

What is the metabolic pathway from hypercholesterolemia to atherosclerosis? On the cellular level, one of the earliest detectable steps in atherogenesis in hypercholesterolemic rabbits and in nonhuman primates is the increased adhesion of monocytes to the endothelium of arteries (Gerrity 1981 a, b; Fagiotto et al. 1984; Fagiotto and Ross 1984), followed by invasion of monocyte/macrophages into the subendothelial space of the intima. Here they are converted to foam cells with large cytoplasmic cholesterolester droplets (Gerrity 1981 a, b; Fagiotto et al. 1984; Fagiotto and Ross 1984).

In 1979 Goldstein et al. first described the so-called scavenger receptor on the surface of macrophages, a high-affinity mechanism for the uptake of modified LDL, in this case acetylated LDL (Ac-LDL), that mediated massive cytoplasmatic accumulation of cholesterolester droplets giving the overloaded cells the appearance of foam cells.

The scavenger receptor has since been shown to recognize a variety of modified LDL with an increased net electronegativity such as Ac-LDL (Goldstein et al. 1979), malondialdehyde-modified LDL (Fogelmann et al. 1980), and apo B-containing particles isolated from atherosclerotic aorta (Goldstein et al. 1981). It also binds maleylated albumin. Some polyanionic compounds such as polyvinylsulfate, polyinosinic acid, and fucoidan are competitive inhibitors of ligand binding to the scavenger receptor (Brown et al. 1980).

Although there is no indication that acetylation of LDL occurs in vivo, other modifications such as oxidation could take place in the circulation, making the modified lipoprotein recognizable by the scavenger receptor. Indeed, LDL incubated with endothelial cells or smooth muscle cells undergoes oxidative modification that results in its recognition by macrophages via the scavenger receptor (Henricksen et al. 1981; Cathcart et al. 1985; Quinn et al. 1985; Parthasaraty et al. 1986; Heinecke et al. 1986). Soutar and Knight (1982) demonstrated high-affinity binding sites for both LDL and modified LDL on normal human monocyte-derived macrophages, whereas in macrophages from patients homozygous for FH only the scavenger receptor was functional.

The scavenger receptor exists on the surface of sinusoidal and Kupffer cells in animals (Nagelkerke et al. 1983; Pitas et al. 1985).

Dresel et al. (1985) showed it to be functional in rat liver in vivo by visualizing selective and rapid uptake of Ac-LDL by sequential scintiscans. Partially purified fractions from rat liver membranes contain two proteins with an apparent molecular weight of 220 kDa and 250 kDa that bind Ac-LDL in a specific and saturable fashion (Dresel et al. 1985). It is assumed that the scavenger receptors's primary physiological function in the liver is the removal of accumulated, deteriorated LDL from the circulation, thus representing a protective system. In contrast, the uptake of modified LDL particles by macrophages in the periphery via the scavenger receptor, leading to foam cell formation, presumably constitutes an early step in atherogenesis. In the development of atherosclerosis due to a cascade of interacting systems, the scavenger receptor is one contributing and possibly integral factor (Steinberg 1983; Ross 1986).

In this paper evidence for the presence of the functional scavenger receptor in the human liver is collected.

Methods

Human LDL (d 1.019–1.063) was isolated from the plasma of normolipidemic volunteers by sequential ultracentrifugation. Acetylation of LDL was performed as described by Basu et al. (1976). Ac-LDL was labeled with ^{125}I according to the iodine monochloride method of Mc Farlane (1958). More than 95% of radioactivity was protein bound and less than 5% discovered in the lipid fraction of the lipoproteins. Lipoproteins were stored after sterile filtration at 4 °C under nitrogen and used within 1 week. Effectivity of acetylation and purity of lipoprotein preparations were checked by agarose electrophoresis. Bovine livers and spleens were purchased from the local slaughterhouse. Human liver and spleen samples were obtained from patients undergoing elective partial hepatectomy or spleenectomy for various reasons (Table 2). Unless dissected and processed immediately, all organ biopsies were transported and stored in liquid nitrogen.

Human liver and spleen samples were carefully examined for malignant or abnormal areas, which were processed separately. Membrane pellets

Table 2. Details of subjects

Subject	Sex	Age (Years)	Reason for Surgery	Serum Cholesterol[a]	
				Total	LDL-Chol
				(mg/dl)	
Membrane binding studies					
SN	F	54	Metastasis of rectum ca[b]	136	99
SE	F	56	Metastasis of rectum ca	159	109
DJ	M	70	Liver cell ca	81	47
SG	F	34	Metastasis of gastric ca	91	62
ZA	F	36	Benign liver cyst	94	61
CN	M	28	Benign liver cyst	78	57
HH	F	51	Metastasis of mammary ca	122	76
Ligand blot					
LJ	M	45	Metastasis of rectal ca	127	84
WJ	M	62	Metastasis of sigmoid ca	183	132
AH	M	68	Metastasis of sigmoid ca	166	118
WA	F	59	Livercell ca	85	74
HH	M	75	Livercell ca	164	99
ME	M	65	Metastasis of rectal ca	116	72
SI	F	48	Metastasis of sigmoid ca	173	99

[a] Determined after 18 h fast at time of surgery.
[b] ca, carcinoma.

were prepared essentially as described previously (Soutar et al. 1986) with membrane buffer (4 ml/g of minced tissue): 50 mM Tris-HCl, pH 8, 150 mM NaCl, 2 mM ethylenediaminetetraacetate (EDTA), 2 mM phenylmethylsulfonyl fluoride. The homogenate was cleared by centrifugation at 1500 g for 10 min at 4 °C and the membrane fraction was prepared from the supernatant by a subsequent ultracentrifugation at 100000 g for 60 min at 4 °C in a Beckmann 60 Ti rotor. Unless processed immediately, membranes were stored in liquid nitrogen until further use.

Binding studies of ^{125}I-labeled Ac-LDL to human liver membranes in the presence and absence of either an excess of unlabeled Ac-LDL or polyvinylsulfate were performed essentially as described (Harders-Spengel et al. 1982). For details see legends to figures.

For solubilization, membranes were resuspended in membrane buffer combined with 1% (v/v) triton X-100 and homogenized in a small Potter glass homogenizer with 20 strokes. After centrifugation at 100000 g for 60 min at 4 °C, the supernatant was diluted fourfold with 50 mM Tris-HCl, pH 8.0, 2 mM Na$_2$EDTA, and 1% (v/v) triton X-100. Detergent exchange from triton X-100 to 40 mM octylglucopyranoside was carried out on small PEI cellulose columns according to Via et al. (1986), followed by elution in a salt step (0.5 M NaCl). The eluates were desalted on Sephadex G-25 equilibrated in 50 mM Tris-HCl, pH 6.8 and stored in liquid nitrogen, if not subjected to sodium dodecyl sulfate polyacrylamide gel electrophoresis (SDS-PAGE) immediately.

The samples were applied to the gels at a final concentration of 1% (w/v) SDS and 10% (w/v) glycerol in the absence of reducing agents (protein contents of samples as indicated in legends to figures). Electrophoretic separation of partially purified receptor preparations was performed on polyacrylamide slab gels in the presence of 0.1% SDS (Laemmli 1970) followed by electrophoretic transfer to nitrocellulose sheets (Towbin et al. 1979). Unless prestained molecular weight marker proteins were used, nitrocellulose strips containing the molecular weight marker proteins were cut off and stained in amido-black.

Nitrocellulose sheets were soaked in 50 mM Tris-HCl, pH 8, 90 mM NaCl, 50 mg/ml bovine serum albumin (BSA) for 60 min at 37 °C to block nonspecific binding sites. After incubation at room temperature for 1 h with the same buffer either with or without the addition of ligand (50 µg/ml Ac-LDL), as indicated in legends to figures, nitrocellulose sheets were washed three times for 20 min in 50 mM Tris-HCl, pH 8, 90 mM NaCl, 5 mg/ml BSA, and subjected to either enzyme-linked immunosorbent assay (ELISA) or incubations for autoradiography, as described in the legends to figures. To test specificity or saturability, competitors or excess unlabeled lipoprotein were included in some incubations as indicated.

Results and Discussion

For membrane binding studies, membrane preparations from macroscopically normal human liver or from metastatic tissue were incubated with increasing concentrations of ^{125}I-Ac-LDL in the presence and absence of either unlabeled Ac-LDL or 50 µg/ml of the polyanionic inhibitor polyvinylsulfate, and the amount of labeled lipoprotein bound was determined.

The amount of Ac-LDL bound increased in a curvilinear manner with increasing ^{125}I-Ac-LDL concentrations, suggesting saturable binding sites for Ac-LDL. The marked inhibition in the presence of excess unlabeled acetyl-LDL (1 mg/ml) in the incubation mixture confirmed that the binding is saturable. The addition of 100 µg/ml polyvinyl sulfate also served to inhibit binding of ^{125}I-Ac-LDL to liver membranes to virtually the same extent (Fig. 1); (Regnström and Harders-Spengel, to be published).

In comparison, membranes from a metastasis in human liver bound ^{125}I-Ac-LDL to a very similar extent at all concentrations (Fig. 2), binding also showing saturability as well as polyvinylsulfate-sensitivity (Regnström and Harders-Spengel, to be published).

K. Harders-Spengel

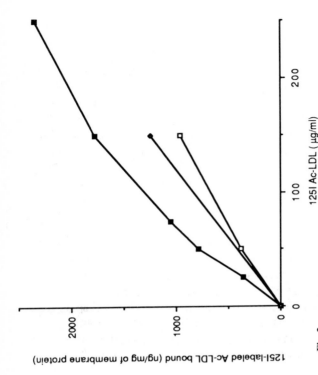

Fig. 1

Fig. 1. Binding of ^{125}I-Ac-LDL to membranes from a normal human liver. Membranes (50 µg in a total volume of 60 µl of buffer) were incubated for 2 h at 0 °C with ^{125}I-Ac-LDL at increasing concentrations, as indicated. The amounts of labeled lipoproteins bound were determined. Binding of ^{125}I-Ac-LDL without further additions = total binding; binding of ^{125}I-Ac-LDL in the presence of an excess of unlabeled Ac-LDL (1 mg/ml); binding of ^{125}I-Ac-LDL in the presence of 100 µg/ml of polyvinylsulfate. Each *point* is the mean of results for triplicate incubations. Specific activity of ^{125}I-Ac-LDL was 82 cpm/ng

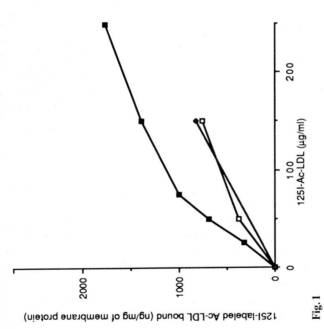

Fig. 2

Fig. 2. Binding of ^{125}I-Ac-LDL to membranes from a metastasis in human liver. Membranes (50 µg in a total volume of 60 µl of buffer) were incubated for 2 h at 0 °C with ^{125}I-Ac-LDL at increasing concentrations, as indicated. The amounts of labeled lipoproteins bound were determined. Binding of ^{125}I-Ac-LDL without further additions = total binding; binding of ^{125}I-Ac-LDL in the presence of an excess of unlabeled Ac-LDL (1 mg/ml); binding of ^{125}I-Ac-LDL in the presence of 100 µg/ml of polyvinylsulfate. Each *point* is the mean of results for triplicate incubations. Specific activity of ^{125}I-Ac-LDL was 82 cpm/ng

34

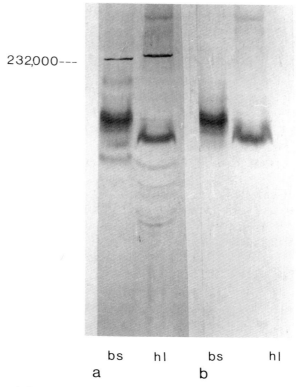

232,000---

bs   hl   bs   hl

a   b

Fig. 3 aq b. Ligand blotting of partially purified fractions from bovine spleen (*bs*) and normal human liver (*hl*). Partially purified protein samples were separated by electrophoresis on a linear 3%–9% acrylamide gradient gel in the presence of 1% (w/v) SDS without reducing agents and electrophoretically transferred to nitrocellulose sheets. For visualization, blots were incubated either with **a** 50 µg/ml of Ac-LDL or **b** without Ac-LDL. The first antibody was linked to the receptor-bound ligands by incubation with rabbit antihuman apo B antiserum (1:1000). The second antibody was introduced by incubation with antirabbit IgG alkaline phosphatase conjugate (1:3000). The blots were developed in a substrate solution of nitro blue tetrazolium/bromochloroindolylphosphate and dimethylformamide in 0.1 M NaHCO$_3$

For visual identification, partially purified fractions from human liver and human spleen were subjected to the ligand blot/ELISA technique after separation of proteins by SDS-PAGE and subsequent electrophoretic transfer to nitrocellulose sheets. The fractions contained a protein with an approximate molecular weight of 232 and 234 kDa, respectively, that bound Ac-LDL (Fig. 3 a).

Sometimes and to varying degrees bands or smears in various lower, and rarely higher, molecular weight areas will be visible. Most of them pre-

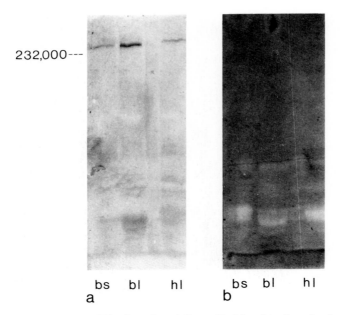

232,000---

bs   bl   hl   bs   bl   hl
a   b

Fig. 4 a, b. Ligand blotting of partially purified fractions from bovine spleen (*bs*), bovine liver (*bl*) and normal human liver (*hl*). Partially purified protein samples were separated by SDS-PAGE (5% acrylamide) in the presence of 1% (w/v) SDS without reducing agents and electrophoretically transferred to nitrocellulose sheets. Blots were incubated with 25 µg/ml of ^{125}I-Ac-LDL either **a** without further additions or **b** in the presence of a 20-fold excess of unlabelled Ac-LDL. For visualization, autoradiography was performed by exposing Kodak XR5 film to blots. The amounts of partially purified protein per lane were bovine spleen, 108 µg; bovine liver, 115 µg; human liver, 120 µg. Specific activity of ^{125}I-Ac-LDL was 179 cpm/ng

sumably bind directly to one of the antibodies or to the enzyme complex employed in the visualization procedure as they also appear when incubations are carried out in the absence of ligand (Fig. 3 b). However, subunits of the scavenger receptor from rat liver at molecular weights of approx. 95, 35, and 15 kDa that specifically bind Ac-LDL have been described by Dresel et al. (1985). In this experiment the presence of such Ac-LDL-binding active subunits can neither be excluded nor proven since no parallel incubation in the presence of a specific inhibitor has been carried out. Therefore the bands at an approx. molecular weight of ≤ 140 kDa that are not visible on the nitrocellulose sheet incubated in the absence of Ac-LDL could represent either subunits of the receptor or nonspecific artifacts.

To test saturability, binding of ^{125}I-labeled Ac-LDL to partially purified membrane proteins separated by SDS-PAGE and transblotted onto nitrocellulose sheets in the presence and absence of an excess of unlabeled

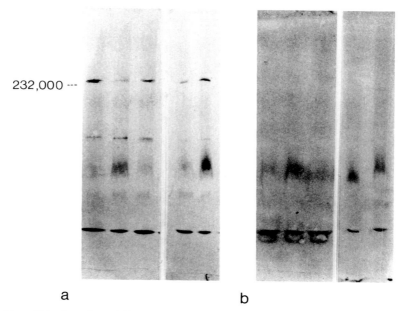

232,000 ---

a   b

Fig. 5 a, b. Ligand blotting of partially purified protein fractions from different human liver samples. Nitrocellulose blots of partially purified protein fractions from five different patients (180 μg of protein/lane) were incubated either with **a** 50 μg/ml Ac-LDL without further additions or **b** with 50 μg/ml Ac-LDL in the presence of 100 μg/ml polyvinylsulfate and developed as described in Fig. 3

Ac-LDL was visualized by autoradiography. Single bands binding ^{125}I-labeled Ac-LDL were visible in partially purified fractions from bovine spleen, bovine liver, and human liver (Fig. 4a), which were abolished by the addition of a twentyfold excess of unlabeled Ac-LDL (Fig. 4b).

Partially purified membrane fractions from five patients were subjected to ligand blot/ELISA after electrophoretic separation on SDS polyacrylamide slab gels and subsequent transfer to nitrocellulose sheets with or without the addition of 100 μg/ml of polyvinyl sulfate to the incubation buffer. The extracts contained proteins with an approx. molecular weight of 232–234 kDa with Ac-LDL binding activity (Fig. 5a). Binding to these proteins was inhibited in the presence of 100 μg/ml polyvinyl sulfate (Fig. 5b). Additional bands of Ac-LDL binding activity at the approx. molecular weight of 145 kDa were also abolished by the addition of polyvinyl sulfate to the incubation buffer. Likewise, there was a certain amount of specific binding visible in the front. They could possibly represent binding-active subunits. Inconstant smears at an approx. molecular weight of 110–116 kDa appeared unchanged after incubation with polyvinyl sulfate and were therefore nonspecific artifacts due to the visualization procedure.

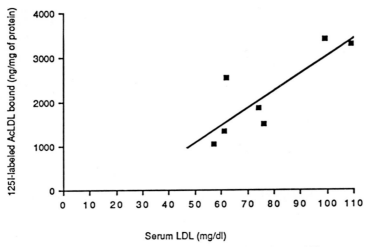

Fig. 6. Relationship between maximal saturable binding of ^{125}I-Ac-LDL to human liver membranes and serum LDL levels. Maximal saturable binding of ^{125}I-Ac-LDL to human liver membranes was determined as described in legends to Figs. 1 and 2. Specific activity of ^{125}I-Ac-LDL was 82 cpm/ng of protein. Each *point* is the mean of triplicate separate determinations. n, 7; r, 0.83; $P < 0.01$. Significance was determined by testing the correlation coefficient r to be different from zero according to R. A. Fischer and F. Yates, Statistical Tables 1963

The addition of maleylated bovine serum albumin (600 µg/ml), fucoidan (100 µg/ml), or polyinosinic acid (100 µg/ml) to the incubation buffer also served to abolish specific binding of Ac-LDL (results not shown). As can be seen in Fig. 5a, there was a marked interindividual variation in the amount of protein specifically binding Ac-LDL in partially purified liver samples from different patients. This phenomenon has been described for the LDL receptor in human liver by Soutar et al. (1986), where the LDL receptor content of human liver varied over a wide range interindividually. Quantifying this finding, Soutar et al. found a strong inverse correlation between the LDL receptor protein content in human liver membranes as determined by radioimmunoassay and plasma LDL cholesterol concentration of the subjects from whom the liver biopsies were obtained ($n = 10$). They hereby confirmed their previous observation of a highly significant inverse relationship between maximal saturable LDL binding to the liver membranes of 15 subjects and their respective plasma cholesterol levels (Harders-Spengel et al. 1982). This correlation is well in keeping with the concept of the regulatory interaction between the LDL receptor system and plasma cholesterol levels.

Testing this approach for the scavenger receptor system in human liver showed, in contrast, a significant positive correlation between maximal saturable binding of Ac-LDL to human liver membranes measured in

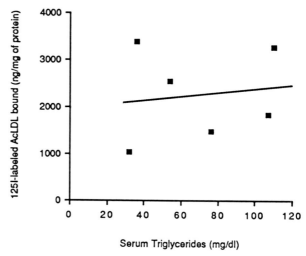

Serum Triglycerides (mg/dl)

Fig. 7. Relationship between maximal saturable binding of ^{125}I-Ac-LDL to human liver membranes and serum triglyceride levels. Maximal saturable binding of ^{125}I-Ac-LDL to human liver membranes was determined as described in legends to Figs. 1 and 2. Specific activity of ^{125}I-Ac-LDL was 82 cpm/ng of protein. Each *point* is the mean of triplicate separate determinations. n, 7; r, 0.15; N.S. Significance was determined by testing the correlation coefficient r to be different from zero according to R. A. Fischer and F. Yates, Statistical Tables 1963

triplicate from seven subjects and their respective serum LDL cholesterol levels (Fig. 6); (Regnström and Harders-Spengel, to be published).

Correlations for various other parameters such as maximal saturable Ac-LDL binding versus serum triglycerides (Fig. 7), serum HDL cholesterol levels, serum VLDL levels, or patients' age did not yield significant trends (Regnström and Harders-Spengel, to be published).

There is as yet no concept for a possible regulatory mechanism for scavenger receptor expression in the liver; therefore, we have to accept the merely descriptive nature of this finding for the time being. Still, this observation fits in with the notion of the scavenger receptor in the liver being responsible for removal of accumulated cytotoxic modified LDL from the circulation, assuming that the height of the LDL level is an indicator for LDL catabolic rate. In spite of the comparatively small number of patient data available for this correlation, the high significance all the more supports its validity. We hope that this trend will be confirmed by our present work on a larger number of human liver biopsies.

The biopsies in this study were taken from a widely heterogeneous group of patients of both sexes, all but two of whom were suffering from malignant and mostly metastasizing disease. Small amounts of malignantly transformed cell material may be present in the liver membrane preparations and in the partially purified receptor fractions. It would be desirable to investigate the questions addressed above in liver samples from normal sex- and age-matched subjects. However, results from limited studies carried out in liver biopsies from the two patients with benign liver cysts (echinococcosis) did not differ from the results obtained from the liver samples originating from patients with malignant disease. Moreover, no qualitative difference between results for metastatic tissue and results for apparently homogeneously normal material could be observed in our studies. We showed saturable and specific binding of Ac-LDL to membranes from liver metastases (Fig. 2), and ligand blot visualization in partially purified fractions from metastases in human liver also revealed a protein that specifically and in a saturable fashion bound Ac-LDL, approx. molecular weight 240–242 kDa (results not shown). Quantitatively, the interindividual variation in apparent scavenger receptor content as judged by the amount of maximal saturable binding to membranes and/or intensity of bands in ligand blot visualization appeared to be even more extreme (Regnström and Harders-Spengel, to be published). LDL receptor activity has been shown to differ dramatically in different malignantly transformed cell types. It is abnormally high in malignant white blood cells (Peterson et al. 1985) and lost in malignantly transformed renal tissue (Clayman et al. 1986). Therefore this preliminary finding is not surprising as the samples studied originated from a selection of morphologically heterogeneous metastases or malignant primary tumors. Further studies need to be carried out for characterization of the scavenger receptor in human liver and to provide data about possible regulatory influences on the expression of the scavenger receptor.

Summary

A specific, saturable binding site for Ac-LDL on human liver membranes has been shown. Partially purified membrane fractions from human liver contain a protein with an approximate molecular weight of 232–234 kDa that binds Ac-LDL in a specific and saturable fashion, frequently accompanied by proteins of lower molecular weight that also exhibit polyvinyl sulfate-sensitive binding of Ac-LDL. Visualization can be achieved by ligand blot, employing ELISA or autoradiography after SDS-PAGE. In partially purified membrane fractions from human spleen, bovine liver, and bovine spleen the same protein can be visualized, the protein from the bovine organs having a slightly higher approx. molecular weight of 240 kDa. In human liver samples from different subjects marked interindividual variation in the amount of protein binding Ac-LDL specifically and in a saturable fashion can be observed both in membrane-binding studies and after visualization by ligand blot. There is a positive correlation between maximal saturable binding of Ac-LDL to human liver membranes from seven persons and their respective serum LDL cholesterol levels. The present work, carried out in the homologous system of human lipoproteins and human organ samples, suggests that the scavenger receptor is expressed and functional in the human liver.

Acknowledgements. I would like to thank my coworkers Bernadette Boisai, Dr. Jan Regnström, Dipl. Biol. Ulrich Reiff, and Dr. Claus Wellhausen for practical assistance and critical discussion. I am indebted to Prof. H. Denecke and the staff of the Chirurgische Klinik, Großhadern, and to Mechthild Haberkamp for help in the collection of liver samples. PD Dr. Martin A. Schreiber, Medizinische Poliklinik der Universität München, provided competent help with the statistical evaluation of data.

References

1. Basu SK, Goldstein JL, Anderson RGW, Brown MS (1976) Proc Natl Acad Sci USA 73:3178–3182
2. Brown MS, Goldstein JL (1986) Science 232:34–47
3. Brown MS, Basu SK, Falck RR, Ho YK, Goldstein JL (1980) J Supramol Struct 13:67–81
4. Cathcart MK, Morel DW, Chisolm GM (1985) J Leukocyte Biol 38:341–350
5. Clayman RV, Bilhartz LE, Spady DK, Buja LM, Dietschy JM (1986) FEBS Lett 196, 1:87–90
6. Dresel HA, Friedrich E, Via DP, Schettler G, Sinn H (1985) EMBO J 4,5:1157–1162
7. Fagiotto A, Ross R (1984) Arteriosclerosis 4:341–356
8. Fagiotto A, Ross R, Harker L (1984) Arteriosclerosis 4:323–340
9. Fogelmann AM, Schechter I, Seager J, Hokom J, Child JS, Edwards PA (1980) Proc Natl Acad Sci USA 76:2214–2218
10. Friedman M, Byers SO, St George S (1962) J Clin Invest 41:828–841
11. Gerrity RG (1981 a) Am J Pathol 103:181–190
12. Gerrity RG (1981 b) Am J Pathol 103:191–200
13. Goldstein JL, Ho YK, Basu SK, Brown MS (1979) Proc Natl Acad Sci USA 76:333–337
14. Goldstein JL, Hoff HF, Ho YK, Basu SK, Brown MS (1981) Arteriosclerosis 1:210–226
15. Harders-Spengel K, Wood CB, Thompson GR, Myant NB, Soutar AK (1982) Proc Natl Acad Sci USA 79:6355–6359
16. Heinecke J, Baker L, Rosen H, Chait A (1986) J Clin invest 77:757–761
17. Henricksen T, Mahoney EM, Steinberg D (1981) Proc Natl Acad Sci USA 78:6499–6503
18. Laemmli UK (1970) Nature 227:680–685
19. McFarlane AS (1958) Nature 182:53
20. Nagelkerke JF, Barto KP, van Berkel TJC (1983) J Biol Chem 258:12221–12227
21. Niehaus C, Nicoll A, Wootton R, Jamieson CW, Lewis J, Lewis B (1977) Lancet II:693–642
22. Parthasarathy S, Printz DJ, Boyd D, Joy L, Steinberg D (1986) Arteriosclerosis 6:505–510
23. Peterson C, Vitols S, Rudling M, Blongren H, Edsmyr S, Skoog L (1985) Med Oncol Tumor Pharmacother 2:143–147

24. Pitas RE, Boyles J, Mahley RW, Bissel DM (1985) J Cell Biol 100:103–117
    Quinn MT, Parthasarathy S, Steinberg D (1985) Proc Natl Acad Sci USA 82:5949–5953
25. Ross R (1986) N Engl J Med 314:488–500
26. Smith EB (1974) Adv Lipid Res, 12:1–49
27. Smith EB (1977) Am J Pathol 86:665–674
28. Smith EB, Slater RS (1972) Lancet I:463–469
29. Smith EB, Staples EM (1980) Proc R Soc Lond [Biol] 217:59–75
30. Soutar AK, Harders-Spengel K, Wade DP, Knight BL (1986) J Biol Chem 261:17127–17133
31. Soutar AK, Knight BL (1982) Biochem J 204:549–556
32. Spengel FA, Jadhav A, Wood C, Thompson GR (1981) Lancet II:768–770
33. Steinberg D (1983) Arteriosclerosis 3:283–301
34. Stender S, Zilversmit DB (1982) Arteriosclerosis 2:493–499
35. Thompson GR, Soutar AK, Spengel FA, Jadhav A, Gavigan SJP, Myant NB (1981) Proc Natl Acad Sci USA 78:2591–2595
36. Towbin H, Staehelin T, Gordon J (1979) Proc Natl Acad Sci USA 76:4350–4354
37. Via DP, Dresel HA, Gotto AM (1986) Methods Enzymol 129:216–226

Expression of Human Apo AI, AII and CII Genes in Pro- and Eukaryotic Cells

W. Stoffel, E. Binczek, A. Haase, and C. Holtfreter

The recent, short-lived boom in cloning of human serum apolipoprotein finally made available cDNA clones and genomic clones of the ten serum lipoproteins. It led to the confirmation of known polypeptide sequences for AI, AII, CI–CIII and the elucidation of unknown primary protein sequences, the most remarkable of which are apo B and apo A. On the basis of such cloning experiments perhaps the functionally and physiologically more important questions may be answered by expression studies of wild type apolipoprotein cDNA or genomic clones and particularly of mutagenized apolipoprotein genes.

This paper reports on studies which provide three selected examples of the enormous potential of molecular biological methods in serum lipoprotein research: in vitro transcription-translation studies with wild type and mutagenized human apo AI; experiments with genomic apo AI and AII DNA and the analysis of expression and secretion products; and the apolipoprotein design used in the molecular biological analysis of the structure–function relationship.

In Vitro Transcription–Translation Studies

The first topic of concerns are in vitro transcription-translation studies with wild type and mutagenized human apo AI. We studied the cotranslational translocation of the preform of apo AI wild type cDNA and of two mutants in which, by site-directed mutagenesis using the gapped duplex method, (a) the peculiar Gln^{-2}-Gln^{-1} C-terminus of the prosequence and the Gln^{-8}-Ala^{-7} C-terminus of the presequence were transposed and (b) the hexapeptide prosequence was deleted.

Among the primary translation products of the serum apolipoprotein RNAs only apo AI and AII carry hexa- and pentapeptide proforms, re-

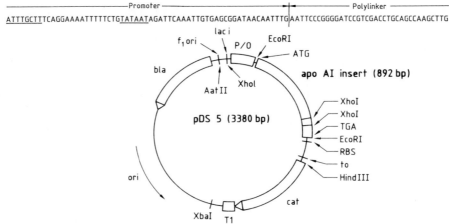

─────────────── Promoter ─────────────── ┼── ──────── Polylinker ────────
ATTTGCTTTCAGGAAAATTTTTCTGTATAATAGATTCAAATTGTGAGCGGATAACAATTTGAATTCCCGGGGATCCGTCGACCTGCAGCCAAGCTTG

Fig. 1. Apo AI/pDS5 clone

Normal sequence (wild type)

	Presequence				Prosequence								
DNA	GGG	AGC	CAG	GCT	CGG	CAT	TTC	TGG	CAG	CAA	GAT	GAA	CCC
	−10			−7	−6					−1	+1		
protein	Gly	Ser	Gln	Ala	Arg	His	Phe	Trp	Gln	Gln	Asp	Gln	Pro

Mutated sequence (mutant 1)

oligonu-	GGG	AGC	CAG	CAG	CGG	CAT	TTC	TGG	CAG	GCT	GAT	GAA	CCC
cleotide	−10			−7	−6					−1	+1		
protein	Gly	Ser	Gln	Gln	Arg	His	Phe	Trp	Gln	Ala	Asp	Gln	Pro

Mutated sequence (mutant 2)

DNA	GGG	AGC	CAG	GCT	– – – – – – – – – – – – – – – –	GAT	GAA	CCC
	−10			−7	−6 −1	+1		
protein	Gly	Ser	Gln	Ala	missing prosequence	Asp	Gln	Pro

Fig. 2. Mutagenized sequences of apo AI

spectively, interposed between signal peptide and mature form. Nothing is known about the function of the propeptide.

Figure 1 schematically represents the expression vector for the wild type and mutant apo AI cDNA, the apo AI insert with 13 bp at the 5′ end, the 801 bp coding region, and 75 bp at the 3′ end, and Fig. 2 the mutations.

The synthetic nucleotides used in gapped duplex mutagenesis are indicated below the mutant preprosequences (Fig. 2). The mutations and the correct orientation of the inserts were confirmed by restriction and DNA sequence analysis (Fig. 3, 4). Transcription-translation in vitro of the wild

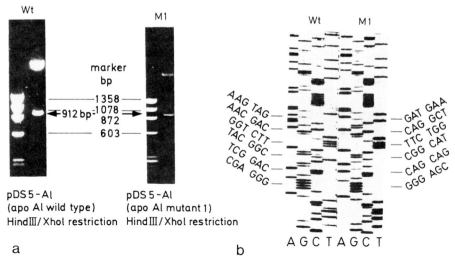

Fig. 3. a *Hin*dIII/*Xho*I restriction analysis of pDS5-AI wild type (*Wt*) and mutant 1 (*M1*). **b** Nucleotide sequence of wild type and mutant 1 of apo AI

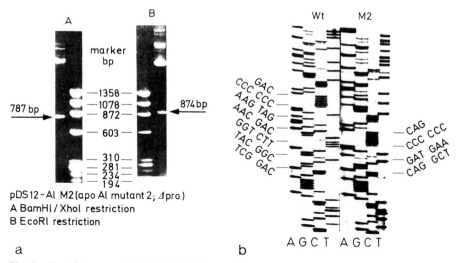

Fig. 4. a Restriction analysis of pDS12-AI mutant 2 with deleted prosequence of apo AI wild type. *A, Bam*HI/*Xho*I restriction; *B, Eco*RI restriction. **b** Nucleotide sequence of apo AI wild type (*Wt*) and apo AI mutant 2 (*M2*) around deleted prosequence

type and of mutant 1 clones combined with translocation and processing by endoplasmic reticulum membranes led to the results summarized in the autoradiograms of Fig. 5.

A 31-kDa primary translation product was synthesized by the wheat germ translation system primed with the in vitro-transcribed apo AI wild-

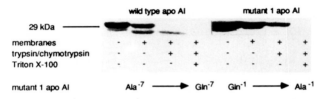

Fig. 5. Comparative in vitro transcription-translation of pDS5-AI wild type and mutant 1 apo AI (Ala^{-7} → Gln^{-7}, Gln^{-1} → Ala^{-1}. 10%–15% NaDodSO$_4$ PAGE of [^{35}S] methionine-labeled polypeptides

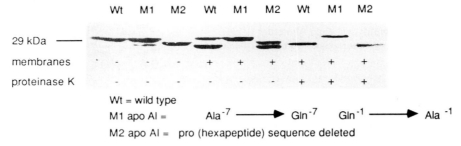

Fig. 6. Comparative in vitro transcription-translation of pDS5-AI wild type apo AI (*Wt*), mutant 1 (*M1*; Ala^{-7} → Gln^{-7}, Gln^{-1} → Ala^{-1}) and mutant 2 (*M2*; prosequence deleted). 10% –15% NaDodSO$_4$ PAGE of [^{35}S]methionine-labeled, immunoprecipitated, apo AI-specific polypeptides

type and mutant-1-specific mRNAs. The wild type primary translation product is translocated cotranslationally and releases the signal peptide by endoproteolysis, yielding an approximately 29-kDA product which is identical with the proform. The mutant 1 primary translation product, however, is not cleaved at all, but is translocated in the luminal side of rabbit or dog endoplasmic reticulum membranes protected against trypsin/chymotrypsin and proteinase K respectively.

When the proform is deleted, as in mutant 2, preapo AI is synthesized, cotranslationally translocated and simultaneously cleaved by signal peptidase to mature apo AI with molecular weight of 28.5 kDa (Fig. 6). The immunoprecipitated methionine- and proline-labeled apo AI forms were radiosequenced by Edman degradation over 30 cycles and the structures confirmed.

From these two site-directed mutagenesis experiments we propose that:

1. The proform is of no relevance for the processing of the signal sequence and therefore contributes neither to a putative linear topogenic se-

quence nor to a topogenic site in a folded tertiary structure around the cleavage site.

2. The sequence following the signal peptide, either the pro- or N-terminal mature sequence, both very different in the case of apo AI, have no influence on the processing.

3. Information is inherent in the signal sequence itself and the correct C-terminus.

The function of the prosequence of apo AI may be different from that of apo AII. In the latter case the deletion leads to cleavage by the signal peptidase within the N-terminus of the mature sequence at the C-terminus of Ala^{+2} instead of Gly^{-1} as shown by Folz and Gordon (1986). Here, however, the accurate cleavage behind the C-terminal Ala^{-1} of the presequence of apo AI is preserved.

We are looking for functions other than that of a simple spacer between signal and mature sequence. The prosequence may be of importance intracellularly for the dichotomy in the posttranslational pathway to a secretory protein, or it may be required intra- and extracellularly for the targeting process required for the assembly with lipids to form the supramolecular structures of the still undefined secreted primary lipoprotein particle.

These mutants, equipped with the regulatory sequences from the genomic clone described below, can now be used for injection experiments in oocytes and to establish transfected cell lines with transient or permanent expression of the apo AI and AII gene. The cell lines allow these and other questions to be addressed.

In Vivo Expression Experiments

In the second part the results of in vivo expression experiments with genomic apo AI and AII DNA and the analysis of expression and secretion products are described.

A fused genomic apo AI – CIII clone of approximately 10 kb in λgt wes was constructed (Fig. 7) and an apo AII clone isolated (Fig. 8). Details of the characterization of both clones and mutants of the apo AII genomic clone for the study of regulatory regions are not shown.

Oocyte Injection Experiments

λAI and AII genomic DNA was injected into the nuclei of *Xenopus laevis* oocytes and the translation products synthesized in the presence of [^{35}S]

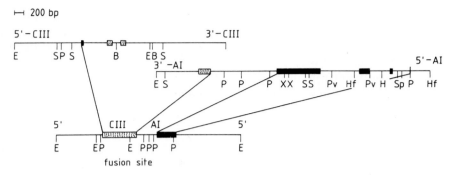

Fig. 7. Construction of fused apo CIII and apo AI genomic DNA with its restriction sites

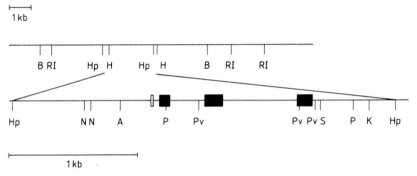

Fig. 8. Apo AII gene organization with restriction sites

methionine analysed (Fig. 9). In addition, [^{14}C] acetate was given as a precursor for labelling lipids of the oocyte newly synthesized in the incubation period and complexed with the secreted apolipoprotein.

The medium of oocytes injected with apo AI–CIII DNA was analysed by CsCl-density gradient centrifugation and the fractions analysed for [^{35}S]methionine-labelled apo AI by immunoprecipitation and by gradient (10%–15%) PAGE and autoradiography and for neutral and complex lipids by radio thin-layer chromatography (TLC).

The results, i.e., that the proforms of apo AI and AII are secreted, were further confirmed by radiosequencing (30 Edman cycles) of [^{3}H]proline- and [^{35}S]cysteine-labelled translation products of AI and AII RNA respectively. Simultaneous injection of apo AI and AII genomic DNA into oocyte nuclei led to the secretion of the two proforms, but they were separated in two densities in the CsCl gradients.

Surprising results appeared in the lipid analysis. The labelling pattern of the oocyte lipids in a TLC radio scan is shown in Fig. 10. No labelled,

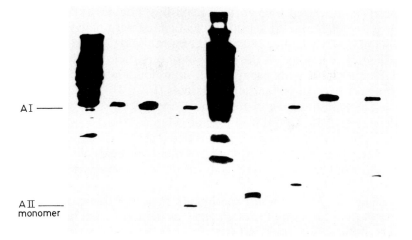

Fig. 9. Translation product synthesized in *Xenopus laevis* oocytes in the presence of [^{35}S]methionine secreted into medium, separated by cell gradient centrifugation and immunoprecipitated with anti-apo AI and AII

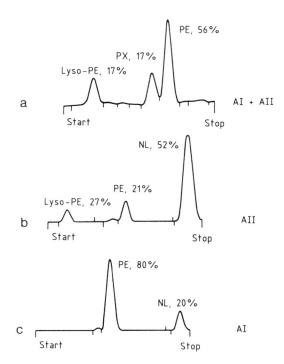

Fig. 10 a–c. Radio thin-layer scans of *Xenopus laevis* oocyte lipids. Oocytes were injected with the DNA listed at the right

newly synthesized lipids are secreted but, oocytes expressing the apo AI gene secreted apo AI selectively associated with newly synthesized phosphatidylethanolamine (80%) and neutral lipid (20%) (Fig. 10 c). Apo AII gene expression led to the secretion of a particle which banded at density 1.10–1.21 g/ml with proapo AII, complexed with phosphatidylethanolamine (PE), lyso-PE and neutral lipids (Fig. 10 b).

The simultaneous expression of apo AI and AII yielded particles of density 1.19–1.21 g/ml with proapo AI and AII associated mainly with PE (56%) lyso-PE (17%) and unidentified Px (17%) (unknown phosphorpositive compound) (Fig. 10 a), and less dense particles of 1.10–1.16 g/ml with proapo AII and exclusively lyso-PE (92%) as newly synthesized lipid, not shown.

The results of these experiments raise the question whether the expression of an apolipoprotein gene, as demonstrated here for apo AI and AII, induces a synthesis of specific phospholipid classes for the formation of a supramolecular structure for secretion which is apolipoprotein and/or cell specific.

Transient and Permanent Expression of Apo AI and AII Genes in CHO and RAT 2 (Fibroblast) Cell Lines

For the study of gene regulation of the main HDL apoproteins AI and AII, not only of the single gene, but also of the interdependence of apo CIII and apo AI gene expression, we have established CHO cell lines with permanent expression. They are dehydrofolate reductase (Dhfr)-negative (auxotroph for thymidine, hypoxanthine and glycine) with the calcium phosphate procedure, and cotransfection with pCVSV-Dhfr cDNA as a marker for transformed cells. Some 80% of the clones contained both DNA species and were selected in methotrexate-supplemented selection medium. Northern blot analysis of CHO cell apo AI- and apo AII-specific RNA confirmed the expression.

The Dhfr gene is a reliable marker for apo AI-transfected CHO cells and RNA dot blots of the clones, but particular ELISAs of the serum-free medium of the cell clones were strongly positive in a dilution of 1 : 100 as compared to the control medium. The apo AI gene was also stably integrated after a period of 2 months.

Furthermore, these cell lines allow experiments, as described here, on regulation of expression in the apo AII gene. The 5′-flanking regions of the apo AII clones in pUC19 were truncated by restriction enzymes HindIII (pAII), NcoI (pAIIΔNco) and ApaI (pAIIΔApa) (Fig. 11). Expression of these DNAs in transiently transfected RAT 2 cells, a rat fibroblast cell line, and CHO cells measured by densitometry of Northern dot-blot hybridization analyses underlines the impact of the 5′-flanking region: λAII with the

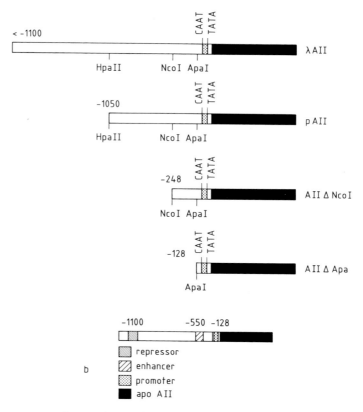

Fig. 11. a 5′ deletion mutants of genomic apo AII DNA. **b** Proposed regulatory elements of 5′-untranslated region of human apo AII gene

> 1100 bp upstream region apparently binds a protein which inhibits splicing and transcription in all cells except liver cells (Hep G2) and enterocytes.

Transcription inhibition is released in pAII which has 1050 bp upstream and pAIIΔNco. The latter still has 240 bp upstream from the transcription start. Deletion by *Apa*I restriction results in a construct which apparently has lost the binding site for a stimulating factor, possibly an enhancer. This and the promoter are tissue-unspecific regulatory sequences. Because of this they are also expressed in only slightly differentiated non-liver cells. Figure 11 b shows a model of the regulation of tissue-specific expression of apo AII by the binding of enhancing and inhibitory proteins. The role of transactivation factors of cellular origin is not yet understood. Present studies are concerned with the properties of the enhancer which are not yet sufficiently understood.

Apolipoprotein Design for Analysis of Structure–Function Relationship

Finally a field of particular interest to us, namely, the apolipoprotein design for the analysis of the structure–function relationship by molecular biological methods, will be described. As a paradigm, apo CII, the activator protein of serum lipoprotein lipase, was studied.

An *Eco*RI insert of a λgt11 apo CII cDNA clone was isolated. Nucleotide sequence analysis indicated that 44 bp were missing at the 5′ end of the coding region. The apo CII clone, 442 bp of full length, was constructed with synthetic oligonucleotides (shown schematically in Fig. 12) and cloned into pUC13.

For expression studies and the analysis of the translation products antibodies against human serum VLDL apo CII and a synthetic peptide embracing Asp^7 to Thr^{16} (KLH conjugate) were raised. The antiserum titre was determined by ELISA. In vitro transcription of the apo CII cDNA cloned into the *Bam*HI/*Pst*I site of the pSP19 vector yielded a 0.4-kb transcript, as shown in the Northern blot with CII 48mer as probe (Fig. 13).

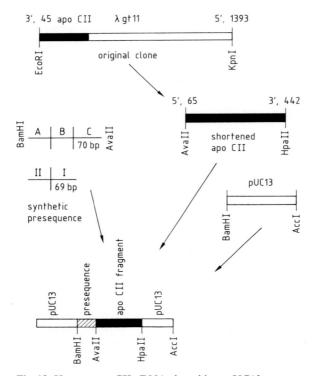

Fig. 12. Human apo CII cDNA cloned into pUC13 vector

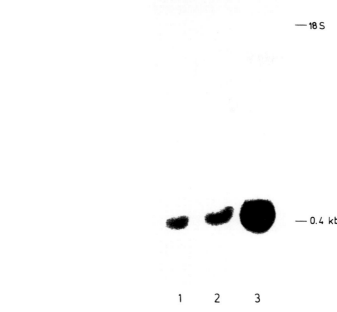

Fig. 13. Northern blot hybridization of *1*, human liver mRNA (OD 40); *2*, human liver mRNA (OD 32); *3*, apo CII transcript from pSP19–CII clone. Probe: apo CII-specific 38mer

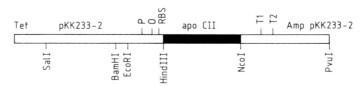

Fig. 14. Human apo CII cDNA cloned into the pKK 233-2 vector

In vitro translation of this apo CII mRNA synthesized by the SP6 RNA polymerase, priming the reticulocyte or the wheat germ system, and purification of the translation product by immunoprecipitation with human apo CII-specific antibodies yielded a 11-kDA protein which is the preform of apo CII with 101 amino acid residues (22 of the pre- and 79 of the mature sequence).

For the production of non-fusion proteins of cloned eukaryotic DNA in bacteria the apo CII cDNA was cloned into the pKK233-2 expression vector (Fig. 14). A sufficient amount of wild type and mutant apo CII was isolated from lysed bacteria with anti-apo CII IgG, purified by apo CII-peptide Affigel 10 affinity chromatography. The medium was free of

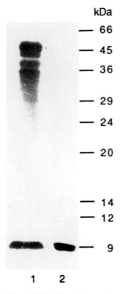

Fig. 15. SDS polyacrylamide gel electrophoresis of apo CII isolated from lysed, *E. coli*-transformed, pKK233-2 apo CII: *1*, pulse-chase experiment; *2*, immunoprecipitated product from lysed cells

apo CII. Apo CII, however, was isolated from the lysed bacteria. It was present as processed, 9-kDA, mature apo CII (Fig. 15).

The laboratory of Gotto has demonstrated that synthetic C-terminal peptides of apo CII can activate lipoprotein lipase (LPL). Other structural

Table 1 clearly demonstrates that the deletion of only six C-terminal amino acid residues (73–79) reduces the activation of LPL to one-third of that of the complete apo CII, and deletion of 14 residues yields an expression product which no longer enhances LPL activity.

Activator	nmol FFA/ml per hour[a]	Factor
Apo CII of VLDL	649	6.4
Apo CII (purified) of bacteria	601	5.8
Apo CII del. 1	209	2.1
Apo CII del. 2	236	2.3
Apo CII del. 3	179	1.8
Apo CII del. 4	147	
Apo CII del. 5	161	
Apo CII del. 6	161	
pKK233-2 wild type lysate (100–400 µl)	75	
Test without activator	102	

[a] Nanomoles of free fatty acid released from triolein/h.

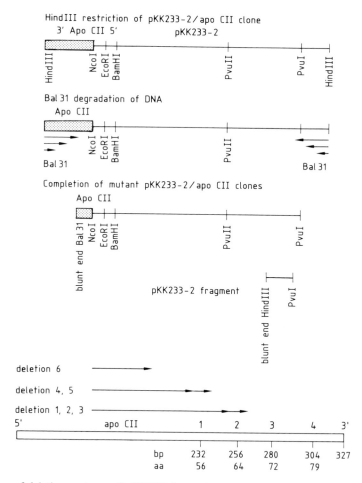

Fig. 16. Construction of deletion mutants of pKK233-2 apo AII with 3'-terminal-truncated coding sequences obtained by partial *Bal*31 exonuclease digestion

and functional domains such as the binding site to LPL and for fatty acids have been postulated. Direct evidence can be derived from 3'-terminal deletions with *Bal*31 exonuclease of the *Hind*III-opened pKK233 apo CII clone (Fig. 16). 24mer oligonucleotides, probing the desired 3'-terminal deletions in six different mutants, were used for selection. These mutants were used for preparative isolation of the C-terminal-truncated apo CII, isolated by affinity immunoadsorption as described above. They were assayed for their LPL activation and compared with apo CII isolated from VLDL.

Conclusion

The potential has been demonstrated of recombinant DNA techniques in studies of structural and functional aspects of lipoprotein research, exemplified in the field of processing and secretion of apo AI, regulation of expression of apo AI and AII in transfected eukaryotic cells and, finally, in the field of apolipoprotein engineering as exemplified for apo CII constructs and their potential for the activation of serum lipoprotein lipase. Future studies open our common field of interest to unexpected and unbelievable horizons.

References

Colman A (1985) Expression of exogenous DNA in *Xenopus* oocytes. In: Hames BD, Higgins SJ (eds) Transcription and translation, a practical approach. IRL, Oxford, pp 49–295

Folz RJ, Gordon JI (1986) Deletion of the propeptide from human preproapolipoprotein A-II redirects cotranslational processing by signal peptidase. J Biol Chem 261:14752–14759

Gurdon JB, Melton DA (1981) Gene transfer in amphibian eggs and oocytes. Annu Rev Genet 15:189–218

Inouye S, Wang S, Sekizawa J, Halegoua S, Inouye M (1977) Amino acid sequence for the peptide extension on the prolipoprotein of the *Escherichia coli* outer membrane. Proc Natl Acad Sci USA 74:1004–1008

Karathanasis SK (1985) Apolipoprotein multigene family: tandem organization of human apolipoprotein AI, CIII, and AIV genes. Proc Natl Acad Sci USA 82:6374–6378

Kramer W, Drutsa V, Jansen HW, Kramer B, Pflugfelder H, Fritz HJ (1984) The gapped duplex DNA approach to oligonucleotide-directed mutation construction. Nucleic Acid Res 12:9441

Kramer W, Schughart K, Fritz H-J (1982) Directed mutagenesis of DNA cloned in filamentous phage: influence of hemimethylated GATC sites on marker recovery from restriction fragments. Nucleic Acid Res 10:6475–6481

Smith LC, Voyta JC, Catapano AL, Kinnunen PKJ, Gotto AM, Sparrow JI (1980) Activation of lipoprotein lipase by synthetic fragments of apolipoprotein CII. Ann NY Acad Sci 2:213–223

Stoffel W, Blobel G, Walter P (1981) Synthesis in vitro and translocation of apolipoprotein AI across microsomal vesicles. Eur J Biochem 120:519–522

Purine Metabolism

Hypoxanthine-Guanine Phosphoribosyltransferase Deficiency: Molecular Basis and Clinical Relevance

B. L. DAVIDSON, T. D. PALELLA, and W. N. KELLEY

Hypoxanthine guanine phosphoribosyltransferase (HPRT) is a cytosolic enzyme which catalyzes the salvage of the purine bases hypoxanthine and guanine to IMP and GMP, respectively. As shown in Fig. 1, the phosphoribosyl moiety is contributed from the substrate, 5-phosphoribosyl-1-pyrophosphate (PRPP).

Deficiency of HPRT activity results in two clinical disorders. Partial HPRT deficiency causes severe hyperuricemia and gout (Kelley et al. 1967). The Lesch-Nyhan syndrome (Seegmiller et al. 1969) results from complete deficiency of HPRT. HPRT deficiency is an X-linked recessive

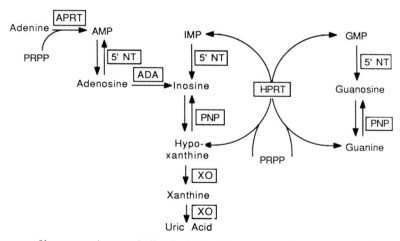

Fig. 1. Summary of human purine metabolism including inborn errors of purine metabolism. HPRT deficiency (reaction 2) results in underutilization of PRPP. Thus, PRPP accumulates and activates de novo purine synthesis (reaction 1). Nucleotides formed by de novo synthesis (guanylic acid and inosinic acid) can not be salvaged in HPRT deficiency and are degraded to uric acid (reaction 6–8). This accounts for the overproduction of uric acid in HPRT deficiency

trait, consistent with the localization of the HPRT gene on the long arm of the X-chromosome (Becker et al. 1979; Pai et al. 1980).

In HPRT-deficient states, PRPP accumulates to levels that activate de novo biosynthesis of purine nucleotides. These nucleotides are ultimately degraded to uric acid. Thus, the clinical consequence of both partial and complete HPRT deficiency is hyperuricemia on the basis of accelerated de novo purine synthesis.

Partial deficiency of HPRT is responsible for hyperuricemia in less than 0.5% of adult subjects with gout. The onset of gout occurs at a young age in most cases, and is often accompanied by uric acid stones. Occasionally, these patients demonstrate mild neurologic dysfunction characterized by hyperreflexia, incoordination, dysarthria, and less frequently, mental retardation (Kelley et al. 1983). B-lymphoblasts derived from patients with HPRT deficiency have levels of enzyme activity ranging from 0.8% to 70% of normal control values, with residual levels of immunoreactive HPRT protein ranging from <0.7% to 70% (Wilson et al. 1986).

Subjects with the Lesch-Nyhan syndrome have hyperuricemia and hyperuricaciduria, and a bizarre neurologic disorder characterized by spas-

Table 1. Properties of HPRT variants

Mutant (patient initials)	Activity[a] (%)	CRM[b] (%)	Kinetic/catalytic defects[c]			Electrophoretic migration[d]
			Km (Hx)	Km (PRPP)	V_{max}	
Gout						
HPRT$_{Ann Arbor}$ (K.C.)	10	4	↑	↑	↑	Cathodal
HPRT$_{Ashville}$ (P.C.)	<0.7	4	↑	↑	N	Cathodal
HPRT$_{London}$ (D.B.)	69	52	↑	N	↑	Normal
HPRT$_{London}$ (G.S.)	59	35	↑	N	↑	Normal
HPRT$_{Munich}$ (I.V.)	3	79	↑	N	↓	Cathodal
HPRT$_{Toronto}$ (L.P.)	23	52	N	N	↓	Anodal
Lesch-Nyhan syndrome						
HPRT$_{Flint}$ (A.C.)	<0.7	<0.5	ND	ND	ND	ND
HPRT$_{Kinston}$ (E.S.)	<0.7	72	↑	↑	N	Cathodal
HPRT$_{Midland}$ (J.H.)	<0.7	<0.5	ND	ND	ND	ND
HPRT$_{Yale}$ (K.T.)	<0.7	92	ND	ND	ND	Cathodal

[a] Percent of control activity in normal B-lymphoblasts. Lower limit of detectability is 0.7% of control.
[b] CRM, cross-reactive material determined by radioimmunoassay. Percent relative to normal B-lymphoblasts. Lower limit of detectability is 0.5% of control.
[c] Michaelis constants (Km) for substrates, hypoxanthine (Hx) and phosphoribosylpyrophosphate (PRPP); V_{max}, maximal initial velocity. ↑, increased; ↓, decreased; N, normal; ND, not determined due to insufficient residual enzyme activity.
[d] Electrophoretic migration in native polyacrylamide gels relative to normal B-lymphoblast HPRT. ND, not determined due to insufficient residual CRM.

ticity, choreathetoid movements, mental retardation, and a compulsive tendency to self-mutilate (Seegmiller et al. 1969; Lesch and Nyhan 1964). The estimated incidence of the Lesch-Nyhan syndrome is 1 per 100 000 live births. The concentrations of HPRT immunoreactive protein in B-lymphoblasts derived from subjects with the Lesch-Nyhan syndrome range from <0.7% to 92%, although most (74%) have undetectable levels (<0.7% of control levels). In the remaining 26%, immunoreactive HPRT protein is present, but enzyme activity is undetectable at normal substrate concentrations (Table 1) (Wilson et al. 1986).

HPRT activity is present in virtually all tissues, but its activity is highest in brain (Rosenbloom et al. 1967). Melton estimated a seven-fold excess of HPRT mRNA in Chinese hamster brain compared to other tissues on the basis of in vitro translation experiments (Melton et al. 1981). Neuronal tissue may be uniquely dependent upon the salvage pathways for purine nucleotides because of the low rate of de novo biosynthesis in these tissues (Watts et al. 1982).

Post-mortem examination of brain tissue from three patients with Lesch-Nyhan syndrome showed significantly decreased indices of dopaminergic function in the caudate nucleus, putamen, pallidum, and nucleus accumbens (Lloyd et al. 1981). In light of the neurologic deficits exhibited in patients with the Lesch-Nyhan syndrome, HPRT activity in brain may be critical for proper neurotransmitter function. At the present time, the exact relationship between HPRT deficiency and neurologic dysfunction remains unclear.

Enzyme Structure

Native HPRT is a tetramer of identical subunits (Holden and Kelley 1978); however, dimers are also active under nonphysiological conditions (Johnson et al. 1979). Each subunit consists of 217 amino acids as determined by peptide sequencing of the purified erythrocyte enzyme, resulting in a monomer molecular weight of 24 470 daltons (Wilson et al. 1982a). The enzyme undergoes post-translational modification that includes the removal of the N-terminal methionine followed by acetylation of the adjacent alanine, and the partial deamidation of asparagine (Wilson et al. 1982a, b). These posttranslational modifications are probably responsible for the electrophoretic heterogeneity of the enzyme demonstrated by isoelectric focusing of erythrocyte lysates and staining for HPRT activity. This heterogeneity increases with enzyme aging.

Structural motifs believed to be involved in the catalytic properties of the enzyme have been postulated (Argos et al. 1983). These regions of amino acid sequence were predicted by comparison of HPRT amino acid sequence with those of other phosphoribosyltransferases. Common secondary structural domains in the phosphoribosyltransferases correlated with the structural features of the nucleotide binding domain of the dehydrogenase family (Argos et al. 1983). The sequences predicted to be necessary for PRPP or hypoxanthine binding are highly conserved from bacteria to rodents and humans (Hershey and Taylor 1986).

Gene Structure

The human HPRT gene is composed of nine exons spread over 44 kilobases (Fig. 2; Patel et al. 1986) and has been localized to the region Xq26–27 between phosphoribosylpyrophosphate synthetase and glucose-6-dehydrogenase (Becker et al. 1979; Pai et al. 1980). The structural coding region encodes a protein of 218 amino acids, which is subsequently cleaved to 217 amino acids. The intron–exon junction sequences of the HPRT gene have been determined and there is total conservation of these boundaries between human and mouse genes (Patel et al. 1986).

   The 5' region of the HPRT gene lacks CAAT and TATA boxes, sequences often associated with the promoters of eukaryotic genes. The immediate 5' region is considerably rich in $G+C$. There are also three copies of the hexanucleotide repeat $^{5'}GGGCGG^{3'}$. Other housekeeping genes such as human and mouse dihydrofolate reductase, human and mouse adenosine deaminase, human 3-phosphoglycerate kinase, and mouse HPRT have copies of this same repeat (Patel et al. 1986; Melton et al. 1986; Farnham and Schimke 1985; Yang et al. 1984a; Valerio et al. 1985; Ingolia et al. 1986; Singer-Sam et al. 1984). Hamster and human 3 hydroxy-3-methylglutaryl coenzyme A reductase have copies of the inverse of this sequence ($^{5'}CCCGCC^{3'}$) as well (Luskey 1987; Osborn et al. 1985). Human and mouse adenine phosphoribosyltransferases have repeats of the inverse sequence only (Hidaka et al. 1987; Dush et al. 1985). Deletion of these sequences leads to loss of promoter activity, similar to the case of the G-C promoter boxes of SV40 virus and herpes thymidine kinase gene (Melton et al. 1986; Everett et al. 1983; McKnight and Kingsbury 1984).

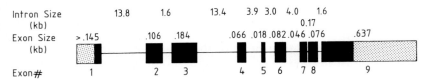

Fig. 2. Structure of the human HPRT gene. The human HPRT spans approximately 44 kb and consists of nine exons (*boxes*) and eight introns (*lines*). *Block regions* indicate coding sequence. *Stippled areas* are transcribed noncoding regions (5'- and 3'-untranslated sequences). The drawing is not to scale

HPRT cDNA sequences isolated from human, hamster, and mouse show a high degree of conservation in both the coding and untranslated regions (>95% and approximately 80%, respectively). Similar to other housekeeping enzymes, HPRT has multiple transcription initiation sites as revealed by primer extension and S1 nuclease studies. The start sites have been mapped from 100 to 170 base pairs 5′ to the translation initiation codon (Patel et al. 1986).

Molecular Basis of HPRT Deficiency

HPRT deficiency may result from: (1) partial or complete gene deletion; (2) defective transcription, leading to low levels or absence of mRNA; (3) unstable, aberrantly processed, or prematurely terminated mRNA; (4) absent or prematurely terminated translation; (5) unstable or rapidly degraded protein; (6) kinetic or catalytic defects; (7) tertiary structural aberrations. Early descriptions of HPRT deficiency suggested that mutations causing this inborn error of metabolism were highly heterogeneous (Seegmiller et al. 1967; Kelley and Meade 1971). Differences existed in levels of enzyme activity, in resistance to product inhibition, in levels of heat stability, in Michaelis constants for the substrates guanine, hypoxanthine, and PRPP, and in electrophoretic migration (Table 1). Subsequent observations have suggested that most mutations in HPRT were point mutations in the structural gene, affecting levels of protein expression rather than message transcription or stability (Wilson et al. 1986).

Different amino acid substitutions in the HPRT protein from three patients with partial HPRT deficiency and one patient with the Lesch-Nyhan syndrome have been identified by peptide sequence analysis (Table 2). HPRT$_{Kinston}$, a patient with Lesch-Nyhan syndrome, retains 70% of HPRT immunoreactive protein, yet the kinetic alterations in this mutant cause an essentially nonfunctional enzyme in vivo. The mutation in HPRT$_{Kinston}$ is an asparagine to aspartate substitution at amino acid 194 (Wilson and Kelley 1983).

HPRT$_{London}$, HPRT$_{Munich}$, and HPRT$_{Toronto}$ have from 35% to 80% of control levels of HPRT protein. A serine to leucine amino acid substitution

Table 2. Amino acid substitutions in HPRT determined by peptide sequencing

Variant	Amino acid substitution		Reference
HPRT$_{Kinston}$	ASP$_{194}$	ARG	Wilson and Kelley (1983)
HPRT$_{London}$	SER$_{110}$	LEU	Wilson et al. (1983b)
HPRT$_{Munich}$	SER$_{104}$	ARG	Wilson and Kelley (1984)
HPRT$_{Toronto}$	ARG$_{51}$	GLY	Wilson et al. (1983a)

was identified in HPRT$_{London}$ at position 110 (Wilson et al. 1983 b). This variant has an elevated Michaelis constant (Km) for hypoxanthine as well as diminished amounts of protein. HPRT$_{Munich}$ also has an elevated Km for hypoxanthine, presumably due to the serine to arginine substitution at amino acid 104 (Wilson and Kelley 1984). HPRT$_{Toronto}$ has a serine to arginine substitution at position 51 (Wilson et al. 1983 a).

Identification of the genetic bases for HPRT deficiency in this heterogenous population has become possible with the isolation and sequencing of HPRT genomic and cDNA sequences and advances in molecular cloning techniques. Southern blot analyses of the genomic DNA from one study of 28 unrelated Lesch-Nyhan patients revealed gross genetic rearrangements of the HPRT gene in 5 (18%) of these patients (Yang et al. 1984 b). These five genetic alterations included two 3' deletions, one total gene deletion, one insertion, and an exon duplication.

A larger survey of HPRT-deficient subjects analyzed for gross genomic alterations indicated that greater than 86% of patients do not have major gene rearrangements. These mutations were shown to be heterogeneous with regard to enzyme activity and levels and kinetic and electrophoretic properties of the enzyme. Approximately normal levels of mRNA were seen in 86% of these patients. Of the 24 unrelated patients in this survey, at least 15 could be identified as unique mutations. Deficient patients possessed insufficient levels of protein for amino acid sequence determination, but retained normal levels of HPRT specific message. This characteristic was exploited in the determination of seven different point mutations in HPRT cDNA clones. cDNA libraries were made using poly(A)$^+$ mRNA isolated from B-lymphoblasts derived from three patients with Lesch-Nyhan syndrome (HPRT$_{Flint}$, HPRT$_{Midland}$, and HPRT$_{Yale}$), and three patients with gout (HPRT$_{London}$, HPRT$_{Ann\ Arbor}$, and HPRT$_{Ashville}$). HPRT-positive clones were isolated and sequenced in their entirety, revealing six point mutations corresponding to five unique amino acid substitutions and one previously determined substitution (Table 3).

Confirmation of these mutations was provided by RNase A mapping techniques or by Southern blotting of genomic DNA. RNase A mapping with HPRT-specific RNA hybridized to RNA from HPRT$_{Flint}$, HPRT$_{Ann\ Arbor}$ and HPRT$_{Ashville}$ yielded cleavage products of expected sizes for the mutations found by cDNA cloning. However, RNase A mapping is useful in approximately 70% of base pair mismatches, depending on the nucleotide context in which these mismatches are located. In the cases of HPRT$_{London}$, HPRT$_{Yale}$, and HPRT$_{Midland}$ RNA, no RNase A cleavage products were obtained.

The mutation in HPRT$_{London}$ predicts the introduction of a new *Hpa*I restriction endonuclease site. To determine whether the amino acid substitution in HPRT$_{London}$ from both patients' (D. B. and G. S.) lymphoblasts are due to the same C to T transition, genomic DNA from both cell lines

Table 3. Nucleotide and predicted amino acid substitutions in mutant human HPRT cDNA clones

Mutant (patient initials)	Mutation[a]		Amino acid substitution[b]		Reference
HPRT$_{Ann\ Arbor}$ (K.C.)	T$_{396}$	G	Ile$_{132}$	Met	Fujimori et al. (1988a)
HPRT$_{Ashville}$ (P.C.)	A$_{602}$	G	Asp$_{201}$	Gly	Davidson et al. (1988a)
HPRT$_{Flint}$ (A.C.)	C$_{222}$	A	Phe$_{74}$	Leu	Davidson et al. (1988b)
HPRT$_{London}$ (D.B.)[c]	C$_{329}$	T	Ser$_{109}$	Asp	Davidson et al. (1988c)
HPRT$_{Midland}$ (J.H.)	T$_{389}$	A	Val$_{130}$	Asp	Davidson et al. (1988d)
HPRT$_{Yale}$ (K.T.)	G$_{211}$	C	Gly$_{71}$	Arg	Fujimori et al. (1988b)

[a] Nucleotide positions are numbered from the A of the AUG translation start codon (position 1) of HPRT mRNA.
[b] Amino acid substitutions are numbered from the initial Met (position 1) of the primary translation product. This residue is post-translationally cleaved to form the native monomer.
[c] Subject D.B. has the identical amino acid substitution as that found in G.S. (HPRT$_{London}$) by amino acid sequencing (see Table 1).

was digested with *Hpa*I and *Hin*dIII and probed with a radioactively labeled HPRT probe. Southern blots of G. S. and D. B. digested DNA demonstrated the presence of an additional *Hpa*I site, confirming the suspected identity between these two mutants (Davidson et al. 1988c).

Further analysis of mutant forms of HPRT will garner new insights into the spectrum of mutations responsible for deficiency of this enzyme. In addition to knowledge gained regarding mechanisms of mutation and human gene structure, these studies provide a basis for understanding critical enzyme structure-function relationships.

Two relatively recent areas of progress greatly facilitate these endeavors. The first is advances in secondary structure prediction and analysis, such as the Chou-Fasman algorithms (Chou and Fasman 1978) and hydrophathicity analyses (Hopp and Woods 1981). These predictive paradigms allow estimates of the impact of altered protein structure on aberrant function. Such first approximations then serve as important guideposts to the design of the more rigorous three-dimensional crystallographic studies required to define structure–function relationships. We have already applied these predictive algorithms to several of the mutants we have defined (Davidson et al. 1988a, 1988d; Fujimori et al. 1988a, 1988b). Although discussion of our results is beyond the scope of this review, our secondary structure predictions have permitted tentative assignment of functions such as substrate binding to certain regions of the HPRT molecule.

The second advance is DNA amplification, using the polymerase chain reaction (Simpson et al. 1988). This approach exploits DNA polymerase to amplify DNA primed with HPRT-specific oligonucleotides. We have reverse transcribed HPRT mRNA isolated from cells derived from affected subjects and then amplified the resultant single-stranded DNA several hundredfold. This mutant DNA may then be sequenced directly or cloned in vectors such as M13 for sequence analysis. We have been able to routinely amplify and analyze HPRT cDNAs obtained in this fashion with lengths exceeding 1 kb (unpublished observation). This dramatic advance in mutant DNA sequence analysis will allow the rapid and accurate definition of mutations which are reflected in mRNA and thus allow the study of many more mutations than otherwise would have been feasible.

The clinical relevance of these studies is still an unexplored frontier, but nonetheless full of promise. A single example which supports this assertion is the consideration of the application of this data to the design of enzymes which exceed the activity and stability of their natural forms. Such engineered proteins could be conveniently encoded into genes installed into vectors under development for potential somatic cell gene therapy.

References

Argos P, Hanei M, Wilson JM, Kelley WN (1983) A possible nucleotide binding domain in the tertiary fold of phosphoribosyltransferases. J Biol Chem 258:6450–6457

Becker MA, Yen R, Itkin R, Gross S, Seegmiller J, Bakay B (1979) Regional localization of the gene for human phosphoribosylpyrophosphate synthetase on the X-chromosome. Science 203:1016–1019

Chou PY, Fasman GD (1978) Empirical predictions of protein conformation. Annu Rev Biochem 47:251–276

Davidson BL, Pashmforoush M, Kelley WN, Palella TD (1988a) Human hypoxanthine guanine phosphoribosyltransferase deficiency: The molecular defect in a patient with gout (HPRT$_{Ashville}$). J Biol Chem (In press)

Davidson BL, Pashmforoush M, Kelley WN, Palella TD (1988b) Genetic basis of hypoxanthine guanine phosphoribosyltransferase deficiency in a patient with the Lesch-Nyhan syndrome (HPRT$_{Flint}$). Gene 63:331–336

Davidson BL, Wilson JM, Kelley WN, Palella TD (1988c) Hypoxanthine guanine phosphoribosyltransferase: genetic evidence for identical mutations in two partially deficient subjects. J Clin Invest (In press)

Davidson BL, Palella TD, Kelley WN (1988d) Human hypoxanthine guanine phosphoribosyltransferase: a single nucleotide substitution in cDNA clones isolated from a patient with Lesch-Nyhan syndrome (HPRT$_{Midland}$). Gene 68:85–92

Dush MK, Sikela JM, Khan SA, Tischfield JA, Stambrook PJ (1985) Nucleotide sequence and organization of the mouse adenine phosphoribosyltransferase gene: presence of a coding region common to animal and bacterial phosphoribosyltransferases that has a variable intron exon arrangement. Proc Natl Acad Sci USA 82:2731–2735

Everett RD, Baty D, Cambon P (1983) The repeated GC-rich motifs upstream from the TATA box are important elements of the SV40 early promoter. Nucleic Acids Res 11:2447–2464

Farnham PJ, Schimke RT (1985) Transcriptional regulation of mouse dihydrofolate reductase in the cell cycle. J Biol Chem 260:7675–7680

Fujimori S, Hidaka Y, Davidson BL, Palella TD, Kelley WN (1988a) Identification of a single nucleotide change in a mutant HPRT gene (HPRT$_{Ann Arbor}$). Hum Genet 79:39–43

Fujimori S, Davidson BL, Kelley WN, Palella TD (1988 b) Identification of a single nucleotide change in the hypoxanthine-guanine phosphoribosyltransferase gene (HPRT$_{Yale}$) responsible for Lesch-Nyhan syndrome. J Clin Invest (In press)

Hershey HV, Taylor MW (1986) Nucleotide sequence and deduced amino acid sequence of *Escherichia coli* adenine phosphoribosyltransferase and comparison with other analogous enzymes. Gene 4:287–293

Hidaka Y, Tarle SA, O'Toole TE, Kelley WN, Palella TD (1987) Nucleotide sequence of the human APRT gene. Nucleic Acids Res 15:9086

Holden JA, Kelley WN (1978) Human hypoxanthine-guanine phosphoribosyltransferase: evidence for a tetrameric structure. J Biol Chem 253:4459–4463

Hopp TD, Woods KR (1981) Prediction of protein antigenic determinants from amino acid sequences. Proc Natl Acad Sci USA 78:3824–3828

Ingolia DE; Al-Ubaidi MR, Yeung C-Y, Bigo HE, Wright DA, Kellems RE (1986) Molecular cloning of the murine adenosine deaminase gene from a genetically enriched source: identification and characterization of the promoter region. Mol Cell Biol 6:4458–4466

Johnson GG, Eisenberg LR, Migeon BR (1979) Human and mouse hypoxanthine-guanine phosphoribosyltransferase: dimers and tetramers. Science 203:174–176

Kelley WN, Meade JL (1971) Studies on hypoxanthine-guanine phosphoribosyltransferase in fibroblasts from patients with the Lesch-Nyhan syndrome: evidence for genetic heterogeneity. J Biol Chem 246:2953–2958

Kelley WN, Wyngaarden JB (1983) Clinical syndromes associated with hypoxanthine-guanine phosphoribosyltransferase deficiency. In: Stanbury JB, Wyngaarden JB (eds) The metabolic basis of inherited disease, 5th ed. McGraw-Hill, New York

Kelley WN, Rosenbloom FM, Henderson JF, Seegmiller JE (1967) A specific enzyme defect in gout associated with overproduction of uric acid. Proc Natl Acad Sci USA 57:1735–1736

Lesch M, Nyhan WL (1964) A familial disorder of uric acid metabolism and central nervous system function. Am J Med 36:561–570

Lloyd KG, Hornykiewicz O, Davidson L, Shannak K, Farley I, Goldstein M, Shibuya M, Kelley WN, Fox I (1981) Biochemical evidence of dysfunction of brain neurotransmitters in the Lesch-Nyhan syndrome. N Engl J Med 305:1106–1111

Luskey KL (1987) Conservation of promoter sequence but not complex intron splicing pattern in human and hamster genes for 3-hydroxy-3-methylglutaryl coenzyme A reductase. Mol Cell Biol 7:1881–189330

Melton DW, Konecki DS, Ledbetter DH, Hejtmancik JF, Caskey CT (1981) *In vitro* translation of hypoxanthine-guanine phosphoribosyltransferase mRNA: characterization of a mouse neuroblastoma cell line that has elevated levels of hypoxanthine guanine phosphoribosyltransferase protein. Proc Natl Acad Sci USA 78:6977–6980

McKnight SL, Kingsbury R (1982) Transcriptional control signals of a eukaryotic protein-coding gene. Science 217:316–324

Melton DW, Konecki DS, Brennand J, Caskey CT (1984) Structure, expression and mutation of the hypoxanthine phosphoribosyltransferase gene. Proc Natl Acad Sci USA 81:2147–2151

Melton DW, McEwan C, McKie AB, Reid AM (1986) Expression of the mouse HPRT gene: deletional analysis of the promoter region of an X-chromosome linked housekeeping gene. Cell 44:319–328

Osborn TF, Goldstein JL, Brown MS (1985) 5′ end of HMG CoA reductase gene contains sequences responsible for cholesterol-mediated inhibition of transcription. Cell 42:203–212

Pai GS, Sprenkle JA, Do T, Mareni CE, Midgeon BR (1980) Localization of loci for hypoxanthine phosphoribosyltransferase and glucose-6-phosphate dihydrogenase and biochemical evidence of nonrandom X-chromosome expression from studies of a human X-autosome translocation. Proc Natl Acad Sci USA 77:2810–2813

Patel PI, Framson PE, Caskey CT, Chinault AC (1986) Fine structure of the human hypo-xanthine phosphoribosyltransferase gene. Mol Cell Biol 6:393–403

Rosenbloom FM, Kelley WN, Miller J, Henderson JF, Seegmiller JE (1967) Inherited dis-order of purine metabolism: correlation between central nervous system dysfunction and biochemical defects. JAMA 202:103–107

Seegmiller JE, Rosenbloom FM, Kelley WN (1967) Enzyme defect associated with a sex-linked human neurological disorder and excessive purine synthesis. Science 155:1682–1684

Simpson D, Crosby RM, Skopik TR (1988) A method for specific cloning and sequencing of human HPRT cDNA for mutation analysis. Biochem Biophys Res Comm 151:487–492

Singer-Sam J, Keith DH, Tani K, Simmer RL, Shively L, Lindsay S, Yoshida A, Riggs AD (1984) Sequence of the promoter region of the gene for human X-linked 3-phosphogly-cerate kinase. Gene 32:409–417

Valerio D, Duyvesteyn MGC, Dekker BMM, Weeda G, Berkvens TM, van der Voorn L, van Ormondt H, van der Eb AJ (1985) Adenosine deaminase: characterization and ex-pression of a gene with a remarkable promoter. EMBO J 4:437–443

Watts RWE, Spellacy E, Gibbs DA, Allsop J, McKeran RD, Slavin GE (1982) Clinical, post-mortem, biochemical and therapeutic observations on the Lesch-Nyhan syndrome with particular reference to the neurological manifestations. Q J Med 201:43–78

Wilson JM, Kelley WN (1983) Molecular basis of hypoxanthineguanine phosphoribosyl-transferase deficiency in a patient with the Lesch-Nyhan syndrome. J Clin Invest 71:1331–1335

Wilson JM, Landa LE, Kobayashi R, Kelley WN (1982b) Human hypoxanthine-guanine phosphoribosyltransferase: tryptic peptides and post-translational modification of the erythrocyte enzyme. J Biol Chem 257:14830–14834

Wilson JM, Kelley WN (1984) Human hypoxanthine-guanine phosphoribosyltransferase: structural alteration in a dysfunctional enzyme variant ($HPRT_{Munich}$) isolated from a pa-tient with gout. J Biol Chem 259:27–30

Wilson JM, Tarr GE, Mahoney WC, Kelley WN (1982a) Human hypoxanthine guanine phosphoribosyltransferase. Complete amino acid sequence of the erythrocyte enzyme. J Biol Chem 257:10978–10985

Wilson JM, Kobayashi R, Fox IH, Kelley WN (1983a) Human hypoxanthine-guanine phos-phoribosyltransferase: molecular abnormality in a mutant form of the enzyme ($HPRT_{Toronto}$). J Biol Chem 258:6458–6460

Wilson JM, Tarr GE, Kelley WN (1983b) Human hypoxanthine (guanine) phosphoribosyl-transferase: an amino acid substitution in a mutant form of the enzyme isolated from a patient with gout. Proc Natl Acad Sci USA 80:870–873

Wilson JM, Stout JT, Palella TD, Davidson BL, Kelley WN, Caskey CT (1986) A molecular survey of hypoxanthine-guanine phosphoribosyltransferase deficiency in man. J Clin In-vest 77:188–195

Yang JK, Masters JN, Attardi G (1984a) Human dihydrofolate reductase gene organization. Extensive conservation of the G+C-rich 5′ non-coding sequence and strong intron size divergence from homologous mammalian genes. J Mol Biol 176:169–187

Yang TP, Patel PI, Chinault AC, Stout JT, Jackson LG, Hildebrand BH, Caskey CT (1984b) Molecular evidence for new mutation at the HPRT locus in Lesch-Nyhan patients. Na-ture 310:412–414

Molecular Basis of AMP Deaminase Isoform Diversity

R. L. Sabina and E. W. Holmes

AMP deaminase (AMP-D) is a ubiquitious enzyme found in all eukaryotic cells. It plays a critical role in energy metabolism in the cell, and it is an integral component of the purine nucleotide cycle. This enzyme and the purine nucleotide are particularly important in skeletal muscle energy metabolism as evidenced by the high activity of AMP-D in skeletal muscle (Lowenstein 1978) and the myopathic state which develops in patients deficient in the activities of AMP-D (Sabina 1988) and S-AMP lyase (Jafken 1984).

Our laboratories have focused on the control of AMP-D expression for several reasons: (1) there are multiple AMP-D isoforms which exhibit tissue-specific and developmental control (Ogasawara 1975); (2) inherited deficiencies of AMP-D are restricted to specific organs, i.e., skeletal muscle (Sabina 1988) and red blood cells (Ogasawara 1982); (3) there is a high frequency of AMP-D deficiency in skeletal muscle, approximately 2% of all muscle biopsies (Sabina 1988); and (4) there are probably multiple etiologies for AMP-D deficiency in skeletal muscle, i.e., primary (inherited) or secondary (acquired as a consequence of an associated neuromuscular disorder; Fishbein 1985).

To understand what is an apparently complex regulatory pathway leading to tissue-specific expression of different AMP-D isoforms and to identify developmental and environmental signals which control isoform expression, we have embarked on a series of studies to develop antibody (Marquetant 1987) and nucleic acid (Sabina 1987) probes for analyzing AMP-D isoform expression. Model systems have been developed which permit us to study AMP-D expression in vivo as well as in cultured cells which can differentiate into myocytes. Rat was selected for these studies because tissues can be easily obtained at different stages of development, and a permanent myogenic cell line, L6, is available for this species of rodent.

AMP-D was purified from rat skeletal muscle and polyclonal antisera produced in rabbits (Marquetant 1987). The protein was cleaved with cyanogen bromide, and peptides were purified for amino acid sequencing. Based on peptide sequences, oligonucleotides were synthesized and a cDNA-encoding AMP-D was cloned from a skeletal muscle library using these oligonucleotides as probes (Sabina 1987). The resultant anti-sera and nucleic acid probes have been used to illucidate AMP-D isoform expression in this rodent model system.

In the adult rat we find that mixed skeletal muscle contains a 2.5 kb transcript (s) and the predominant AMP-D peptide has a subunit molecular weight of 80 kDa (Sabina 1987). This mRNA and peptide are not detectable in nonmuscle tissues such as brain and kidney, nor are they produced in heart, another striated muscle. This transcript and peptide increase more than 10-fold in abundance from birth of the animal to the development of mature skeletal muscle. Dissection of skeletal muscle into slow-twitch and fast-twitch fibers reveals that the abundance of this transcript and peptide are three times greater in fast than slow twitch fibers. We conclude from these studies that a 2.5 kb AMP-D mRNA (or mRNAs) is produced in skeletal, striated muscle fibers, and this transcript encodes an 80 kDa AMP-D peptide. This transcript is apparently restricted to striated skeletal muscle since it is not detectable in cardiac striated muscle or nonmuscle tissues. The abundance of this transcript is regulated developmentally during muscle maturation, and its abundance is differentially regulated in different fiber types. We conclude from these observations that the expression of a skeletal muscle transcript of AMP-D responds to tissue-specific and developmental signals during muscle formation and maturation.

During intrauterine development, two other isoforms of AMP-D are expressed in rat skeletal muscle (Marquetant 1987). Prior to 15 days of gestation the predominant AMP-D isoform is a 78 kDa peptide, which we have referred to as the embryonic isoform in accord with the temporal expression of other muscle-specific genes such as myosin heavy chain. This isoform is produced in all embryonic tissues, and it is also found in many, but not all, nonmuscle tissues of the adult. From approximately 15 days of gestation up to birth (21 days) the predominant isoform in hind limb is an AMP-D peptide with a subunit molecular weight of 77.5 kDa. We have referred to this peptide as the perinatal isoform. The perinatal isoform is not detectable in nonmuscle tissues at any stage of development, but it continues to be expressed to varying extent in different fiber types of the adult animal. Peptide mapping studies indicate that the 78 kDa embryonic, 77.5 kDa perinatal, and 80 kDa adult isoforms of AMP-D are distinct peptides with unique primary amino acid sequences. In addition, they have distinct kinetic, immunologic, and physical properties. We conclude from these studies that muscle development is characterized by the sequential expression of at least three AMP-D isoforms. These AMP-D isoforms apparently have unique amino acid sequences, suggesting that different AMP-D mRNAs encode these different isoforms. Based on these studies we predict that other AMP-D cDNAs, in addition to the one we have sequenced (Sabina 1987), will be found.

There are multiple mechanisms which could account for the production of different AMP-D mRNAs and peptides. These additional transcripts could arise from more than one AMP-D gene; and/or alternative transcription start sites, alternative exon splicing, or alternative polyadenylation signals could lead to the production of more than one mRNA from a single primary transcript of one gene. Cloning of these other cDNAs and their cognate gene(s) will be required to completely understand the molecular basis of AMP-D isoform diversity. Additional studies are also needed to understand the tissue-specific and stage-specific cis-acting and trans-acting factors which control AMP-D expression during myocyte development. As a preliminary step to these studies we have shown that rat L6 myoblasts can be induced to differentiate in vitro and recapitulate the sequential expression of embryonic, perinatal, and adult isoforms of AMP-D, similar to what we observe in vivo. In addition we have developed differentiation-defective mutants of L6 cells arrested at the "embryonic" and "perinatal" stages of AMP-D expression. These cells should prove useful in dissecting out the control mechanisms which regulate AMP-D isoform switching during muscle development.

The availability of molecular probes and model systems should make it possible to understand both primary and secondary deficiencies of AMP-D at the molecular level in humans. It is likely that humans go through a similar stage- and tissue-specific expression of different isoforms of AMP-D (Kaletha 1988). Thus, AMP-D deficiency is likely to have multiple etiologies, and more than one molecular mechanism will undoubtedly be found to explain AMP-D deficiency in different disorders and in different tissues. Given that there may be other AMP-D isoforms in addition to the ones discussed above and some of these are likely to exhibit tissue specificity, in the future we may find new disorders which are the result of AMP-D deficiency restricted to these organs.

References

1. Fishbein W (1985) Myoadenylate deaminase deficiency: inherited and acquired forms. Biochem Med 33:158
2. Jafken J, VandenBergh G (1984) An infantile autistic syndrome characterized by the presence of succinylpurines in body fluids. Lancet :1058
3. Kaletha K et al. (1987) Developmental forms, human skeletal muscle AMP deaminase. Experienta 43:440–443

4. Kaletha K, Nowak G (1988) Developmental forms of human skeletal muscle AMP deaminase: the kinetic and regulatory properties of the enzyme. Biochem J 249:255
5. Marquetant R et al. (1987) Evidence for sequential expression of three AMP deaminase isoforms during skeletal muscle development. Proc Natl Acad Sci USA 84:2345
6. Ogasawara N et al. (1975) Isoenzymes of rat AMP deaminase. Biochim Biophys Acta 403:530
7. Ogasawara N et al. (1984) Complete deficiency of AMP deaminase in human erythrocytes. Biochem Biophys Res Comm 122:1344–1349
8. Sabina R et al. (1987) Cloning and sequence of rat myoadenylate deaminase cDNA: evidence for tissue-specific and developmental regulation. J Biol Chem 262:12397
9. Sabina R et al. (1988) Myoadenylate deaminas deficiency. Metabolic Basis of Inherited Disease, 6th edn. McGraw-Hill, New York
10. Lowenstein JM, Goodman MN (1978) Purine nucleotide cycle in skeletal muscle. Fed Proc 37:2308

Purine Excretion

R. Greger

Introduction

In many mammals possessing the enzyme uricase purine is excreted mostly as allantoin (Table 1). In primates, including humans, which have lost the ability to convert urate to allantoin, uric acid/urate is the excreted substance (Greger et al. 1975). Two problems connected to the loss of uricase and to the exclusive excretion of uric acid and urate have developed in primates and, more specifically, humans: The first is the limited solubility of

Table 1. Renal excretion and metabolism of urate in different species. (Modified from Lang 1981)

Species	P_{urate}	C_{urate}	Condition resulting in net secretion of urate	Uricase
Man	250	0.1	Renal insufficiency, diuretics, sulfamerazine + mannitol	Negative
Apes	120–300	0.1	Mersalyl	Negative
Old World monkey	30	1.0	Control	Positive
New World monkey	30–200	0.05–0.1	Control	Weak
Pig	6	3.0	Control	Positive
Dalmatian dog	60	0.9	Mannitol infusion	Weak
Mongrel dog	30	0.2–0.9	Mannitol and urate infusion	Positive
Guinea pig	90	0.6	Mannitol and urate infusion	Positive
Rabbit	12– 30	0.1–2.0	Urate infusion	Positive
Rat	60–100	0.2–0.5	Mannitol infusion	Positive
Cat	?	0.6–1.0	Urate infusion	Positive
Goat	20– 60	1.3–3.8	Urate infusion	Positive
Calf	35	1.2	Urate infusion	Positive

P_{urate}, plasma urate concentration in μmol/l; C_{urate}, renal urate clearance as a fraction of glomerular filtration rate.

urate in body fluids, leading to gout; the other is the limited solubility of urate / uric acid in urine, leading to urolithiasis (Mertz 1975). Both phenomena have triggered research in the fields of purine biochemistry and renal uric acid excretion (Holmes, this volume; Kelley, this volume; Lang 1981).

Over the years it has been discomforting that no adequate model of renal urate transport has been available. Several species, including the rat and dog, exhibit substantial net reabsorption of urate and in this respect are comparable to man, although the endogenous plasma water concentrations are only on the order of 50 µmol/l in rat, but 5 times as high in man. Others, including the rabbit, the dalmatian dog, and the guinea pig show net secretion of urate (Lang 1981). Beyond that it has been very discouraging that within one species, i.e., the rat, the data obtained in various laboratories have shown discrepancies to a large extent (Kahn and Weinman 1985; Lang 1981). It has been argued that this was due to serious artifacts in some, but not in other laboratories. However, as pointed out several years ago (Lang 1981), the diversity of data is more likely due to the complex mechanisms of urate transport and the even more complex control. This article focuses only on a few more recent reports and addresses the topic in the form of but a few statements. No attempt will be made to review the recent excellent reviews by Weiner (1979), Lang (1981), and Kahn and Weinman (1985).

Lack of uricase

The lack of uricase generates the problems. Purine metabolism produces urate and in most mammals this is converted (mainly in the liver) to allantoin. As an end product of metabolism allantoin has important advantages over urate. It is excreted effectively in the mammalian kidney, inasmuch as it is only filtered, not reabsorbed to any large extent (Greger et al. 1975). As a consequence, a large quantity of allantoin can be excreted in urine. The other advantage of allantoin is its high water solubility. Thus, problems occuring with uric acid, such as gout or uric acid urolithiasis, are unlikely to materialize. As noted above, primates have largely lost the ability to convert urate to allantoin, and, therefore, are confronted with these problems.

Bidirectional transport

Man is one of those species in which urate is reabsorbed by the kidney. As shown in Table 1, in humans the amount of urate excreted in urine falls short of the amount filtered. In fact, usually only some 10% of the filtered load is excreted in urine; the remaining 90% of the filtered load is reabsorbed. Up to about 10 years ago researchers claimed that urate transport in this group, and in humans specifically, comprised proximal reabsorption and distal secretion. This view is certainly inappropriate, inasmuch as urate secretion and reabsorption occur in the same nephron segment (Lang et al. 1972). In fact, this complex phenomenon of proximal bidirectional transport has made it difficult to analyze the cellular mechanisms of urate transport (Lang 1981). It is also apparent from Table 1 that in humans, as in this entire group, net secretion of urate can be observed under very extreme pathological or experimental conditions.

Bidirectional proximal tubule urate transport involves anion carriers in both the apical (luminal) and the basolateral cell membrane. Previous studies using microperfusion, micropuncture and microinjection have indicated that both reabsorption and secretion are mediated events, but it has been difficult to describe these processes kinetically since the affinity towards urate is rather low (Lang et al. 1972). Recent reports on the cellular mechanisms of urate are based on membrane vesicle studies (Guggino et al. 1983; Kahn and Weinman 1985; Kahn et al. 1985) or peritubular uptake studies (Grantham et al. 1987; Ullrich and Rumrich 1988). The latter method was refined only a few years ago and has been utilized to study the peritubular uptake of a variety of substances, including weak organic acids (Ullrich et al. 1987). The former method (vesicle technique) was designed for the characterization of transport phenomena in purified plasma membrane preparations under well-defined experimental conditions. Recent data unambigously indicate that urate is taken up by the luminal membrane (brush border membrane) via an anion exchanger which seems to operate electroneutrally and exchanges urate for another organic anion or for OH^- (Guggino et al. 1983; Kahn and Weinman 1985). It is unclear which anion is exchanged physiologically for urate. The claim that OH^- or HCO_3^- may be the predominating anion (Kahn and Weinman 1985) seems unlikely since cytosolic HCO_3^- or OH^- concentrations are probably lower than those in the luminal fluid. A very intriguing possibility has been suggested by Kahn and Weinman (1985) and Guggino and Aronson (1985), namely, that urate uptake across the apical membrane may be tertiarily active in that the same membrane takes up organic anions via sodium-coupled cotransport systems. These organic anions then accumulate in the cytosol and can be recycled via sodium-independent urate/anion exchange. Another feature of this urate/anion exchanger is its sensitivity towards known inhibitors of anion transport such as the stilbene derivatives SITS or DIDS (Kahn and Weinman 1985). Also, drugs such as probenecid and even loop diuretics in high doses can inhibit this system. The peritubular transport mechanism has not been studied as well, probably because the preparation of pure basolateral membrane vesicles is more difficult than that of apical membranes. In a recent report (Kahn et al. 1985) the transport is described as anion exchange, however, with properties distinct from those of the apical membrane urate/anion exchanger. For example, the system appears to be less sensitive to inhibitors such as DIDS or probenecid. Furthermore, it was claimed that this system, unlike the luminal transport system was not influenced by paraaminohippurate. On the other hand, in experiments with the intact kidney and a stopped flow technique to measure peritubular uptake processes the different anion transport systems have been well characterized (Ullrich and Rumrich 1988), and it was found that urate and paraaminohippurate compete for the same transport system. This transport system, because of its preferential acceptance of paraaminohippurate, has been labeled the PAH system. In another study utilizing basolateral membrane vesicles it has been shown that this PAH transport system is again effectively tertiarily active since it utilizes other anions (dicarboxylates) as counteranions which are taken up in a sodium-dependent manner (Shimada et al. 1987). Thus, it appears likely that urate moves in this anion transport system, probably in an outward direction to account for urate reabsorption, and probably in an inward direction like PAH to account for urate secretion.

Paradoxical drug effects

Paradoxical drug effects can be explained by differential interference with luminal and basolateral urate/anion carriers or by inhibition as opposed to transstimulation of a single type of exchanger. Paradoxical drug effects on urate clearance have been reported for various drugs. At one dose range the substance is uricosuric, at another dose antiuricosuric. Examples of drugs with paradoxical effects are summarized in a masterpiece table (Lange 1981), quoting some 350 original reports. Salicylate, sulfonamides, thiazides, loop diuretics of the furosemide type, PAH, and pyrazinoate belong to this group. Such puzzling effects could be explained if the drug under study inhibited at some dose the one and at a higher dose the other urate/anion exchanger. An alternative explanation has been given recently for pyrazinoate (Guggino and Aronson 1985). It was shown that urate uptake into brush border membrane vesicles under Na^+ gradient conditions was stimulated by low and inhibited by high concentrations of pyrazinoate. The data were interpreted to indidate that pyrazinoate at low doses was taken up into the membrane vesicles by a Na^+-coupled process and that it transstimulated urate uptake. At higher doses urate uptake was inhibited at the urate binding site.

The table on factors influencing urate clearance quoted above (Lang 1981) can be continued as new, potentially important regulators of urate transport are reported (Grantham et al. 1987). Plasma proteins appear to inhibit urate uptake across the basolateral membrane. The physiological or pathophysiological relevance of this finding is entirely open.

Proximal tubule transport

Urate transport is probably confined to the proximal nephron. It has already been stated that urate is not secreted in the distal nephron, but that secretion and reabsorption occur in the same, namely, the proximal nephron segment. Micropuncture and microperfusion studies in the rat (Lang 1981) have illustrated that urate transport occurs mainly in the proximal convolution and to a limited extent in the segment between late superficial proximal tubule and early distal tubule. Reabsorption in the loop (Greger et al. 1974) as in the proximal convolution is flow dependent: fractional reabsorption is high at low flow rates and lower at high flow rates. This mechanism is probably responsible for the uricosuric effect of loop diuretics (Lang et al. 1977). There is no convincing evidence that any urate transport occurs in the distal nephron.

The pyrazinoate test

The pyrazinoate test is not a reliable indication of the ability to secrete urate. The underlying concept of this test has been that pyrazinoate selectively inhibits urate secretion and that the corresponding decrease in urate clearance is a measure of the ability of the kidney to secrete urate. This concept is probably not applicable quantitatively since reabsorption and secretion of urate occur simultaneously. Thus, much of the secreted urate is reabsorbed and the pyrazinoate test by necessity will underestimate urate secretion (Lang 1981). Furthermore, it was stated above that pyrazinoate not only inhibits secretion, but that it also inhibits urate reabsorption (Guggino and Aronson 1985).

Renal hyperuricemia

The cause of hyperuricemia in humans may lie in the kidney. The easiest way to distinguish individuals who overproduce from those who underexcrete urate is by measuring daily urate excretion and the urate clearance (Lang et al. 1980). Normal individuals will excrete a daily urate load of 2–3 mmol. Overproducers will excrete more. Hyperuricemic patients with normal excretion, but reduced urate clearance can be labeled underexcreters. In these individuals uricosuric drugs increase the urate clearance and normalize plasma urate concentration.

Renal insufficiency

Renal insufficiency may lead to reduced fractional reabsorption of urate and a loss of sensitivity towards uricosuric drugs. In renal insufficiency the flow rate in the remaining intact nephrons is increased, which reduces the fractional urate reabsorption in these nephrons (Greger et al. 1974). This phenomenon helps to prevent early development of hyperuricemia in renal insufficiency. On the other hand, these individuals will respond poorly to uricosuric drugs (Lang et al. 1980).

Conclusion

The present short review has addressed only a few issues pertinent to the general topic. Currently, a coherent overview of the detailed cellular mechanisms of urate transport in the kidney cannot be given. Even though important new insights have been gained recently about urate transfer across cell membranes, the processes are very complex in that they involve tertiarily active transport and thus are modulated not only by the Na^+ gradient and its regulation, but also by the availability of various other organic anions and the regulation of their transport. Furthermore, drug effects are very difficult to interpret since the same drug can act as counterion and/or inhibitor on one or both transport systems of the proximal tubule cell. When it comes to the human kidney, the current situation is probably even less well defined since there are no data from direct studies and since extrapolation from other species appears problematic. Much more work on the various anion transport systems is needed before we have a comprehensive view of how urate moves across the proximal tubule. Recent systematic work on the various anion transport systems (Ullrich and Rumrich 1988) is certainly a step in this direction.

References

Grantham JJ, Kennedy J, Cowley B (1987) Tubule urate and PAH transport: sensitivity and specificity of serum protein inhibition. Am J Physiol 252:F683–F690
Greger R, Lang F, Deetjen P (1974) Urate handling by the rat kidney. IV. Reabsorption in the loops of Henle. Pflugers Arch 352:115–120

Greger R, Lang F, Deetjen P (1975) Handling of allantoin by the rat kidney. Clearance and micropuncture data. Pflugers Arch 357:201–207

Guggino SE, Aronson PS (1985) Paradoxical effects of pyrazinoate and nictinate on urate transport in dog renal microvillus membranes. J Clin Invest 76:543–547

Guggino SE, Martin GJ, Aronson PS (1983) Specificity and modes of the anion exchanger in dog renal microvillus membranes. Am J Physiol 244:F612–F621

Kahn AM, Weinman EJ (1985) Urate transport in the proximal tubule: in vivo and vesicle studies. Am J Physiol 249:F789–F798

Kahn AM, Shelat H, Weinman EJ (1985) Urate and p-aminohippurate transport in rat basolateral vesicles. Am J Physiol 249:F654–F661

Lang F (1981) Renal handling of urate. In: Greger R, Lang F, Silbernagl S (eds) Renal transport of organic substances. Springer, Berlin Heidelberg New York, pp 234–261

Lang F, Greger R, Deetjen P (1972) Handling of uric acid by the rat kidney. II. Microperfusion studies on bidirectional transport of uric acid in the proximal tubule. Pflugers Arch 335:257–265

Lang F, Greger R, Deetjen P (1977) Effects of diuretics on uric acid metabolism and excretion. In: Siegenthaler W, Beckerhoff R, Vetter W (eds) Diuretics in research and clinics. Thieme, Stuttgart, pp 213–224

Lang F, Greger R, Oberleithner H, Griss E, Lang K, Pastner D (1980) Renal handling of urate in healthy man in hyperuricemia and renal insufficiency: circadian fluctuation, effect of water diuresis and of uricosuric agents. Eur J Clin Invest 10:285–292

Mertz DP (ed) (1975) Gicht. Thieme, Stuttgart

Shimada H, Moeves B, Burckhardt G (1987) Indirect coupling of Na$^+$ of p-aminohippuric acid uptake into rat renal basolateral membrane vesicles. Am J Physiol 253:F795–F801

Ullrich KJ, Rumrich G (1988) Contraluminal transport systems in the proximal renal tubule involved in secretion of organic anions. Am J Physiol 254:F453–F462

Ullrich KJ, Rumrich G, Klöss S (1987) Contraluminal paraaminohippurate transport in the proximal tubule of the rat kidney. III. Specificity: monocarboxylic acids. Pflugers Arch 409:547–554

Weiner IM (1979) Urate transport in the nephron. Am J Physiol 237:F85–F92

Nephrolithiasis in the Context of Purine Metabolism

H. A. SIMMONDS

Introduction

Germany has long been in the forefront in highlighting the importance of purines to humans. It is virtually the centenary of the production of the correct structure for uric acid by the great Emil Fischer and his school (1884) and their demonstration of the relationship with other purine bases. The first dietary purine loading experiments were performed in the next two decades (Minkowski 1898) when it was recognised that a significant amount of purine was still excreted on a purine-free diet and the distinction between endogenous and exogenous purine was proposed by Burian and Schuur (1900).

Professor Zöllner (1960) has continued this tradition and made a special contribution in underlining not only the importance of diet, but the specific purine content of the diet as a risk factor for hyperuricaemia and gout. The purine laboratory in Munich will also go down in history for devising the purine-free formula diet and demonstrating that 7 days are required to reach a steady state (Loffler et al. 1981). Like others here today it was an interest in diet – in my case curiosity about the effect of dietary purines and allopurinol on purine excretion (Simmonds et al. 1973), using the only suitable animal model the pig – that started a long and enjoyable link with the purine laboratory in Munich directed by Professor Zöllner (Zöllner and Gröbner 1970). The description of guanine gout in pigs due to a reputed lack of the enzyme guanase stimulated my own experiments. These showed that pigs did not lack guanase and led me through the early work done in Germany, back to Virchow – the only man who said they did – in 1866. The importance of the experiments was the awareness engendered of the superb work from this country as well as the nephrotoxicity of purine bases due to their extreme insolubility, and of course the fact that diet can play a crucial role in this (Simmonds et al. 1973).

H. A. Simmonds

Purines Associated with Nephrolithiasis

The three endogenous purine bases associated with nephrolithiasis in man
are uric acid, xanthine and 2,8-dihydroxyadenine (2,8-DHA). All three are
metabolic end products (Fig. 1). Uric acid is the normal end product of pu-

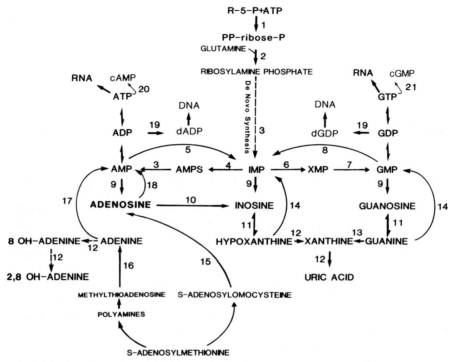

Fig. 1. The purine metabolic pathways in man. The different purine pathway enzymes are
indicated as follows:

1 phosphoribosylpyrophosphate synthetase (EC 2.7.6.1);
2 amidophosphoribosyltransferase
(EC 2.4.2.14);
3 adenylosuccinate lyase (EC 4.3.2.2);
4 adenylosuccinate (AMPS) synthetase
(EC 6.3.4.4);
5 AMP deaminase (EC 3.5.4.6);
6 IMP dehydrogenase (EC 1.2.1.14);
7 GMP synthetase (EC 6.3.5.2);
8 GMP reductase (EC 1.6.6.8);
9 5′-nucleotidase (EC 3.1.3.5);
10 adenosine deaminase (EC 3.5.4.4);
11 purine nucleoside phosphorylase
(EC 2.4.2.1);

12 xanthine oxidase (EC 1.2.3.2);
13 guanine deaminase (EC 3.5.4.3);
14 hypoxanthine – guanine phosphoribosyltransferase (EC 2.4.2.8);
15 S-adenosylhomocysteine hydrolase
(EC 3.3.1.1);
16 methylthioadenosine phosphorylase
(EC 2.4.2.28);
17 adenine phosphoribosyltransferase
(EC 2.4.2.7);
18 adenosine kinase (EC 2.7.1.20);
19 ribonucleotide reductase (EC 1.17.4.1);
20 adenylate cyclase (EC 4.6.1.1);
21 guanylate cyclase (EC 4.6.1.2)

Table 1. Solubility of different purines in human urine

pH	5.0		8.0	
	mg %	mmol/l	mg %	mmol/l
Uric acid	15	0.9	200	12.0
Xanthine	8	0.5	13	0.9
Hypoxanthine	140	10.3	150	11.0
2,8-Dihydroxyadenine	0.3	0.02	0.5	0.03
Oxipurinol	20	1.32	70	4.6

rine metabolism excreted in the urine. Xanthine and 2,8-DHA do not accumulate except in special circumstances discussed below. Consequently, uric acid stones are the most prevalent and have been known to mankind for thousands of years, as exemplified by their having been found in Egyptian mummies. By contrast xanthine stones were first identified by Marcet in 1817 and 2,8-DHA stones were first reported independently just over a decade ago by Cartier et al. in France and ourselves in the UK (Simmonds et al. 1988). Recognition of 2,8-DHA as the real component in the stones from a child with supposed "uric acid stones" was facilitated in my case by knowledge of the nephrotoxicity of 2,8-DHA demonstrated by the early German workers following the feeding of adenine to animals (Minkowski 1898).

The potential nephrotoxicity of these three purines relates to their poor solubility within the normal pH range of human urine (Table 1). At pH 5.0 the solubility is less than 1 mmol/l, and while the solubility of uric acid may be increased twelvefold by alkalinisation, that of xanthine and particularly 2,8-DHA is little changed. Any factor(s) effectively decreasing the solubility of these purines in urine could produce crystalluria and nephrolithiasis.

Aetiology of Purine Nephrolithiasis

Factors resulting in overproduction or overexcretion of uric acid may be endogenous: due to inherited, acquired, or associated metabolic disorders, or exogenous, due to diet/drugs. Factors reducing solubility may also be endogenous, such as metabolic defects or may be climatic or clinical in nature.

Primary Causes of Purine Overproduction:
Genetic Metabolic Defects

Hypoxanthine-Guanine Phosphoribosyltransferase (HGPRT) Deficiency

Professor Kelley has described the molecular mechanisms resulting in uric acid overproduction in the X-linked defect HGPRT deficiency and the clinical associates of this (see this volume). HGPRT deficiency leads to a block in the salvage of hypoxanthine and guanine (Fig. 1), resulting in deficient nucleotide feedback control as well as the accumulation of phosphoribosylpyrophosphate (PP-ribose-P) which further accelerates purine synthesis de novo. The neurological manifestations depend on whether the genetic alteration of the enzyme allows significant HGPRT activity under physiological conditions.

In the context of nephrolithiasis both types of defect, the complete type presenting in childhood or the partial type generally presenting later in adolescence, are associated with gross purine overproduction and hyperuricaemia. Urinary uric acid levels of up to 10 mmol (1600 mg/day) may be recorded which can lead to urolithiasis and frequently acute renal failure. A third of the patients we have studied have presented in this way. Heterozygous females are difficult to detect, but some may show increased uric acid excretion.

PP-ribose-P Synthetase Superactivity

Patients with the other X-linked purine disorder associated with uric acid lithiasis, PP-ribose-P synthetase superactivity, by contrast have normal HGPRT activity. PP-ribose-P synthetase catalyses the reaction of ribose-5-phosphate with ATP to form PP-ribose-P, an allosteric regulator of purine synthesis de novo (Fig. 1). The enzyme requires Mg^{2+} and inorganic phosphate as an obligatory activator and is normally inhibited by ADP, GDP and other nucleotides. The uric acid overproduction in this case is not due to lack of nucleotide inhibitors, but a variety of molecular defects, both catalytic and regulatory, leading to reduced sensitivity to feedback inhibition and overactivity under physiological conditions (Becker et al. 1988).

The defect has now been demonstrated in some 20 families. Hyperuricaemia and uric acid excretion of the same order as in HGPRT deficiency is frequently recorded. Purine overproduction and urolithiasis are generally also present in the heterozygous female. An association between the kinetic defect and the severity of the phenotypic expression, as in HGPRT deficiency, has been suggested from the finding of a family with metabolic and neurologic abnormalities intermediate between the majority who presented in adolesence with gout and urolithiasis, compared with the four families presenting in infancy with the combined syndrome of purine over-production, inherited sensorineural deafness. Resolution of the putative association between these two will require characterisation of the genetic basis for the superactivity and detailed studies of the PP-ribose-P synthetase gene.

Adenine Phosphoribosyltransferase (APRT) Deficiency

APRT normally catalyses the salvage reaction in which adenine condenses with PP-ribose-P in the presence of Mg^{2+} to form AMP (Simmonds et al. 1988). In subjects with a genetic defect in this enzyme, adenine accumulates and is oxidised by xanthine oxidase (Fig. 1) via the 8-hydroxy intermediate to 2,8-dihydroxyadenine (2,8-DHA). As indicated above 2,8-DHA is very insoluble and the clinical associates of this disorder are crystalluria, renal stones in 85% of patients and acute renal failure in up to one-third of these. Two-thirds have been children.

Two types of defect have now been identified in patients in whom 2,8-DHA stones form, depending on the level of residual APRT in erythrocyte lysates. Those with virtually undetectable enzyme activity (type I) – found predominantly in Caucasians, in 35 patients from 12 countries to date, including one from Germany (Jung et al. 1987) – are homozygotes or compound heterozygotes for the null alleles designated APRT*QO. Those with approximately 25% of normal APRT lysate activity (type II) – more than 33 patients exclusively from Japan – are considered to have a mutant allele designated APRT*J. A single mutation appears to be responsible for this defect in Japanese people.

The mutant enzyme from the Japanese patients with the type II defect shows a reduced affinity for PP-ribose-P as well as altered kinetics and heat stability; the K_m for PP-ribose-P is increased tenfold. Since heterozygotes for the type I defect also have approximately 25% of normal activity, homozygotes for the type II defect can only be distinguished by intact cell studies.

APRT is coded by alleles at a gene locus on chromosome 16. The defect shows an autosomal recessive mode of inheritance and has also been the focus of considerable interest at the molecular level. The normal human APRT gene has been cloned and sequenced. Mutant APRT genes for the type 1 defect have recently been isolated, characterised and sequenced, revealing different mutations in the nucleotide sequence on each allele (Hikada et al. 1987).

Xanthine oxidase Deficiency

This genetic defect produces a block in the normal oxidation of hypoxanthine to xanthine and xanthine to uric acid (Fig. 1). The block is virtually complete, with xanthine and to a lesser extent hypoxanthine replacing uric acid as the purine end product excreted in the urine. Of cases presented 30% showed symptoms related to the defect, urolithiasis or xanthine nephropathy, but in the majority the defect has been an incidental finding during investigation of an unrelated clinical problem. More than 100 cases are now known.

Combined xanthine oxidase/sulphite oxidase deficiency has recently been described in several children in which xanthinuria is accompanied by the severe neurological dysfunction associated with sulphite oxidase deficiency. The defect results from the absence of a common cofactor containing molybdenum and appears to relate to inability to synthesise the pteridyl moiety of the cofactor.

Genetic heterogeneity was demonstrated in the defect by the ability of some, but not all, patients with xanthinuria to convert allopurinol to oxipurinol. Workers in Munich have now made a significant contribution to our understanding of this phenomenon by demonstrating that this heterogeneity relates to the presence or absence of aldehyde oxidase activity (Reiter et al. 1988). The locus of the gene coding for xanthine oxidase has not been defined, but inheritance is autosomal recessive.

Secondary Causes of Purine Nephrolithiasis

Purine Overproduction: Genetic and Acquired

Genetic. The glycogen storage disease glucose-6-phosphatase deficiency is an example of an unrelated genetic defect that can have a secondary effect on purine metabolism. A combined effect resulting in increased purine biosynthesis de novo has been postulated to account for the hyperuricaemia and hyperuricosuria seen in this defect. The hypothesis is that the phosphate consumed during glycogen breakdown in response to hypoglycaemia would release controls on adenylate deaminase, thereby stimulating ATP breakdown and uric acid production, in turn releasing phosphoribosylamidotransferase from nucleotide inhibition (Wyngaarden and Kelley 1983). The rapid rise in uric acid levels following the administration of fructose has been ascribed to a similar mechanism.

Diet. A diet rich in purines can be a precipitating factor in the development of purine nephrolithiasis in normal individuals as well as any of the above disorders. "... Fermented drinks are likely to be rich in purines derived from yeast. There is said to be as much purine in a litre of Munich beer as there is in 100 grams of beef." (Peters and Van Slyke 1946). High intake of adenine-rich foods by one APRT-deficient patient may have played an important part in the development of the renal lesion, progression to dialysis and eventually transplantation (Simmonds et al. 1988). As Professor Zöllner and his co-workers have shown (1977), dietary purines can double plasma and urine uric acid levels. Although not relevant to this review, it should be noted that there is also evidence for an aetiological role for hyperuricosuria in calcium oxalate stone formation. This is an area of considerable dispute, but it is possible that dietary purine restriction, or allopurinol therapy, may be of value in some patients in whom oxalate stones form. Trace elements may also be important in that xanthine stones have been reported in sheep grazing on molybdenum-deficient pastures (Easterfield and Bruce 1930), and an acquired deficiency of molybdenum has produced a similar situation in patient on parenteral nutrition (Abumrad et al. 1979).

Reduced Tubular Reabsorption of Uric Acid

Hereditary Renal Hypouricaemia. In this case hypouricaemia, not hyperuricaemia, is associated with increased urinary uric acid excretion. This tubular reabsorption defect has been reported in approximately 33 patients and is inherited as an autosomal recessive trait (Gaspar et al. 1986). The fractional excretion of uric acid via the kidney is greater than the norm (which approximates 10% of the glomerular filtration rate, GFR) and ranges from 20% of the GFR to levels in excess of the GFR. The response to pharmacological inhibitors has varied and there appear to be a variety of mutations where reabsorption of filtered or secreted urate may be defective or secretion even enhanced. The incidence of urolithiasis is generally low, except in families where renal hypouricaemia is also associated with hypercalciuria.

Other Tubular Defects. A similar situation pertains in Wilson's disease, Hartnup disease, Fanconi's syndrome and renal tubular acidosis. All are also associated with increased renal clearance of uric acid by the kidney.

Drugs and Diet. The enhancement of uric acid clearance by different pharmacological agents such as probenecid and benzbromarone has been exploited in the treatment of gout, but adequate measures are usually taken to avoid stone formation (Zöllner et al. 1970). Acute uric acid nephropathy and sometimes urolithiasis, however, have been reported following therapy with a variety of uricosuric drugs which include sulphinpyrazone and radiocontrast agents. Vitamin C is also uricosuric. The uricosuric effect of a high protein diet is perhaps not as widely recognised, but was first demonstrated by early research and has been recently extended by Professor Zöllners group (Loffler et al. 1981).

Iatrogenic Lithiasis

Allopurinol therapy in the treatment hyperuricaemia may lead to xanthine nephropathy and xanthine lithiasis, either in inherited disorders associated with uric acid overproduction, or gross uric acid overproduction as occurs during agressive therapy for different malignancies (Cameron and Simmonds 1987). Oxipurinol stones have also been reported in rare cases in patients treated with allopurinol. Adenine, too, used as therapy to reduce purine synthesis in HGPRT deficiency has produced 2,8-DHA nephrotoxicity.

The experimental models of xanthine nephropathy produced in the pig by administering large doses allopurinol together with dietary purine in the form of guanine and that of uric acid and 2,8-DHA nephropathy in rats fed large doses of uric acid plus oxonic acid, or adenine respectively, have shown that intratubular deposition of these purines can cause interstitial scarring and permanent renal damage as a long-term result (de Vries and Sperling 1977).

Nephrolithiasis Associated with Factors Reducing Solubility

Genetic Defects Associated with a Low Urinary pH

Idiopathic Uric Acid Nephrolithiasis. In idiopathic uric acid nephrolithiasis plasma and urine uric levels are normal or even low, but a low urine pH between 4.8–5.0 is a consistent finding (de Vries and Sperling 1977). Urine is invariably saturated with uric acid when the pH is below 5.5, hence the urolithiasis. An inherited defect was described first in Israel, but the number of cases worldwide is low.

Gout. The prevalence of stones in patients with gout is much higher than in the general population and ranges from 10%–25% (Wyngaarden and Kelley 1983). Gouty patients also have a tendency toward unusually acidic urine, both while fasting and throughout the day, and a substandard rise in response to alkali. Diet may also be a contributing factor since in the majority of gouty patients uric acid excretion is normal on a controlled purine intake (Zöllner et al. 1977).

Excessive Extrarenal Water Loss

Climate. Uric acid stones account for up to 5% of all stones in most countries except in arid climates and in countries fringing the Mediterranean where they can constitute up to 50% of the total (de Vries and Sperling 1977). Climate is also a significant factor in the expression of xanthine nephrolithiasis, which is relatively uncommon in Northern Europe, but frequent in countries bordering the Mediterranean, and has led to nephrectomy in patients from the Middle East. However, geographical location does not seem to be an important factor in the aetiology of 2,8-DHA lithiasis which has been found with equal frequency in temperate and arid climates, as exemplified by the fact that 4 cases have now been reported in Iceland (Simmonds et al. 1988).

Clinical. Patients with chronic diarrhoea or ileostomy for inflammatory bowel disease may also excrete low urine volumes, also with a low pH. The prevalence of stone formation in this group ranges from 7%–12%, demonstrating that dehydration can be an important precipitating factor in purine nephrolithiasis.

Diagnosis

All three types of stone are radiolucent. If stones or gravel have been passed, appearance and physical properties may give some clue as to their identity. Uric acid stones are generally yellowish, hard, smooth and crush with difficulty, whereas xanthine stones are generally orange-brown, smooth and oval, have a laminated appearance when cut and crush more readily. 2,8-DHA stones are generally putty coloured and crush with ease (Simmonds et al. 1988).

Laboratory diagnosis without recourse to modern technology can be misleading, in that all three types of stone give a similar reaction in the classic murexide test or thermogravimetric analysis and in some colorimetric tests. 2,8-DHA stones were invariably mistaken for uric acid in the past. If sophisticated equipment is not available, 2,8-DHA can be distinguished from uric acid by resistance to uricase, xanthine by conversion to uric acid by xanthine oxidase. The three stone types can be distinguished readily by high performance liquid chromatography (HPLC) as well as UV, infrared or mass spectrometry.

Treatment

As in all types of urolithiasis the most effective treatment is first to ensure an adequate water intake. Diet is equally important. Patients should be encouraged to avoid purine-rich foods, and drugs aggravating purine excretion should not be administered.

Alkali is effective only for uric acid lithiasis and is contraindicated for 2,8-DHA lithiasis. Allopurinol – with the dose carefully adjusted in the presence of either purine overproduction or renal disease – is effective in uric acid as well as 2,8-DHA lithiasis (Cameron and Simmonds 1987).

Summary

Uric acid stones are the most common purine stones (Table 2). Five inherited purine disorders can lead to uric acid lithiasis: HGPRT deficiency, PP-ribose-P synthetase superactivity, gout, familial idiopathic urolithiasis

Table 2. Factors predisposing to purine nephrolithiasis

Endogenous	Stone composition		
	Uric acid	Xanthine	2,8-DHA
Primary			
Specific enzyme defect			
HGPRT −	+		
PP-ribose-P Synth	+		
XOD −		+	
XOD/SOD −		+	
APRT −			+
Undefined molecular defect			
Tubular defect	+		
Idiopathic lithiasis	+		
Gout	+		
Secondary			
Glucose 6-phosphatase −	+		
Other tubular defects	+		
Malignancies	+		
Exogenous			
Diet			
Purine: yeast/beer/offal	+		
Protein	+		
Molybdenum deficiency		+	
Adenine			+
Therapy			
Allopurinol therapy		+	
X-ray contrast agents	+		
Ileostomy	+		
Adenine-CPD blood			+

−, Indicates deficiency of a particular enzyme.

and hereditary renal hypouricaemia. APRT deficiency (types I and II) is associated with 2,8-DHA stones; xanthine oxidase deficiency with xanthine stones. Diet, drugs, as well as other inherited or acquired disorders can mimic or exacerbate the genetic defects; dehydration can be a precipitating factor. pH is important only for uric acid stones.

References

Abumrad NN, Schneider AJ, Steele DR, Rogers LS (1979) Acquired molybdenum deficiency. Clin Res 27:774A

Becker MA, Puig JG, Mateos FA, Jiminez ML, Kim M, Simmonds HA (1988) Inherited superactivity of phosphoribosylpyrophosphate synthetase: association of uric acid overproduction and sensorineural deafness. Am J Med (In press)

Burian R, Schur H (1900) Über die Stellung der Purinkörper im menschlichen Stoffwechsel. Drei Untersuchungen. Pflugers Arch 80:241

Burian R, Schur H (1910) Über die Stellung der Purinkörper im menschlichen Stoffwechsel. Drei Untersuchungen. Pflugers Arch 187:239

Cameron JS, Simmonds HA (1987) Use and abuse of allopurinol. Br Med J 294:1504–1505

de Vries A, Sperling O (1977) Implications of disorders of pruine metabolism for the kidney and urinary tract. In: Purine and pyrimidine metabolism in man. Cib Found Symp 48:179–206

Easterfield TE, Bruce A (1930) The occurrence of xanthine calculi in New Zealand Sheep. N Z J Sci Technol 11:357–361

Fischer E (1884) Über die Harnsäure I. Ber Dtsch Chem Ges 17:828–838

Gaspar GS, Puig JG, Mateos FA, Cabanillas AJ (1986) Hypouricaemia due to renal tubular defect. Arch Intern Med 146:1241–1243

Hikada Y, Palella TH, O'Toole T, Tarle SA, Keley WN (1987) Human adenine phosphoribosyltransferase: identification of allelic mutations at the nucleotide level as a cause of complete deficiency of the enzyme. J Clin Invest 80:1409–1415

Jung P, Brommert R, Jesberger H-J (1987) 2,8-Dihydroxyadenine lithiasis: a case with a complete deficiency of adenine phosphoribosyltransferase. Klin Wochenschr 65 (Suppl X):12–13

Loffler W, Gröbner W, Zöllner N (1981) Nutrition and uric acid metabolism: plasma level, turnover, excretion. Fortschr Urol Nephrol 16:8–18

Minkowski O (1898) Untersuchungen zur Physiologie und Pathologie der Harnsäure bei Säugethieren. Arch Exp Pathol Pharmacol 41:375–420

Peters JP, Van Slyke DD (1946) Quant Clin Chem

Reiter S, Simmonds HA, Zöllner N, Braun S, Knebel M (1988) On the role of aldehyde oxidase in the metabolism of allopurinol: demonstration of the combined deficiency of xanthine oxidase and aldehyde oxidase in xanthinuric patients not forming oxipurinol (Submitted)

Simmonds HA, Rising TJ, Cadenhead A, Hatfield PJ, Jones AS, Cameron JS (1973) Radioisotope studies of purine metabolism in the pig during the administration of guanine and allopurinol in the pig. Biochem Pharmacol 22:2553–25563

Simmonds HA, Sahota AS, Van Acker KJ (1989) Adenine phosphoribosyltransferase deficiency and 2,8-dihydroxyadenine lithiasis. In: Scriver CR, Beaudet AL, Sly WS, Valle D (eds) The metabolic basis of inherited disease, 6th edn. McGraw-Hill, New York, Chapt 39 (In press)

Virchow R (1866) Die Guanin-Gicht der Schweine. Virchows Arch [A] 36:147–148

Wyngaarden JB, Kelley WN (1983) Gout. In: Stanbury JB, Wyngaarden JB, Fredrickson DS, Goldstein JL, Brown MS (eds) The metabolic basis of inherited disease, 5th edn. McGraw-Hill, New York, pp 1043–1115

Zöllner N (1960) Moderne Gichtprobleme. Ätiologie, Pathogenese, Klinik. Ergeb Inn Med Kinderheilkd 14:231–389

Zöllner N, Gröbner W (1970) Effects of allopurinol on endogenous and exogenous urates. Eur J Clin Pharmacol 3:56–58

Zöllner N, Gröbner W (1977) Dietary feedback regulation of purine and pyrimidine biosynthesis in man. In: Purine and pyrimidine metabolism in man. Cib Found Symp 48:165–179

Zöllner N, Dofel W, Gröbner W (1970) Die Wirkung von Benzbromaron of die renale Harnsäureausscheidung Gesunder. Klin Wochenschr 48:426–432

Therapeutic Aspects

Hyperlipidemia

Prognosis of Dyslipoproteinemia

P. Schwandt

The prognosis of a disease depends on several variables such as age at manifestation, spontaneous course of the disease, treatment possibilities, therapy risks, and accompanying diseases. This is illustrated by the history of a girl aged 6 years and 9 months with severe heart disease secondary to homozygous familial hypercholesterolemia. The spontaneous course of this genetic disorder manifested early in life proved to be treatable by liver transplantation, which meant the transfer of functional LDL receptors with a consequent decrease in cholesterol levels. However, the risk of this operation was very high, especially because a heart transplantation had to be done simultaneously (Starzl et al. 1984; Bilheimer et al. 1984). Nevertheless, several years after the operation the cholesterol level is near to normal (assisted by additional treatment with a lipid-lowering drug) and the young patient is well.

Combined Risk Factors of Atherosclerosis

The prognosis of atherosclerosis depends furthermore on additional disorders and risk factors for coronary heart disease (CHD). These have been divided into modifiable risk factors and other factors in the Policy Statement of the European Atherosclerosis Society (Study Group European Atherosclerosis Society 1988; Table 1).

The combination of several risk factors results in a steep increase in mortality due to coronary disease. The Framingham data clearly show that the range of serum cholesterol between 221 to 244 mg/dl resulted in 3.8 deaths due to coronary disease per 1000 men (35 to 57 years of age) within 6 years if they were normotensive and did not smoke; if they were hypertensive and smokers the death rate was elevated to 16.6 within the same cholesterol range (Mc Gee 1973).

P. Schwandt

Table 1. Factors affecting therapy of hyperlipidemia

Modifiable risk factors	Other factors
Hypertension	Family history of coronary heart disease or peripheral vascular disease
Cigarette smoking	Personal history of early onset of coronary heart disease
Diabetes mellitus	Revascularization procedures
Obesity	Male sex
Low HDL cholesterol	Younger age

The same is true for the combined effects of high density lipoprotein (HDL) cholesterol and low density lipoprotein (LDL) cholesterol on the relative risk of CHD (data are taken from a 24-year follow-up of the Framingham Study in men aged 50 to 70 years). The relative risk in men with LDL cholesterol concentrations of 160 mg/dl was 1.5 when HDL cholesterol (25 mg/dl) was low. It was decreased to 0.2 when HDL cholesterol levels (85 mg/dl) were high (Kannel et al. 1979).

Familial Hypercholesterolemia

The dyslipoproteinemia with the worst prognosis is the homozygous form of familial hypercholesterolemia (FH) as demonstrated by the above case report. The marked elevations of LDL cholesterol from birth lead to the very early development of severe CHD (the youngest patient with myocardial infarction reported was 18 months old). Very rarely do patients become older than 30 years.

The heterozygous form with an incidence of 1/500 is caused by deletion of one LDL receptor gene. Though the prognosis is better than in the ho-

Table 2. Incidence and age distribution of CHD in more than 1000 patients heterozygous for FH

Age (years)	CHD symptoms (%)		Deaths due to CHD (%)	
	m	f	m	f
40	20	3	0	0
50	45	20	25	2
60	70	45	50	15
70	90	75	80	30

m, male; f, female.

mozygous form, these patients also have severe CHD which already becomes manifest in middle age (Table 2).

Familial Dysbetalipoproteinemia

In contrast to the genetic defect of the LDL receptor in FH familial dysbetalipoproteinemia (type III hyperlipoproteinemia) is due to a defect of the ligand: the apolipoprotein E_2 (apo E_2) has a cysteine residue instead of an arginine in position 158 of the polypeptide chain. This structural defect in the apo E_2 molecule inhibits the binding of apo E-containing lipoprotein particles to the apo B, E receptor. The consequence of this is an increase of apo E_2-containing remnants of chylomicrons and of very low density lipoprotein (VLDL) which give rise to a high incidence of severe atherosclerosis occuring early in life. To cause type III hyperlipoproteinemia (with an incidence of 1/10000), besides the E_2/E_2 phenotype (incidence 1/100) an additional factor is needed (for instance another form of Familial hyperlipoproteinemie hypothyroidism, or diabetes mellitus). In a group of 185 patients with type III hyperlipoproteinemia the incidence of CHD was 28% and that of peripheral arterial disease was 21% (Janus et al. 1985). As in FH the prognosis of this disease depends on the possibility of normalizing the severe atherogenetic dyslipoproteinemia with combined dietary and drug treatment. This therapy should be initiated as early as possible.

The Chylomicronemia Syndrome

Extremely high chylomicron concentrations in plasma acutely affect the prognosis of genetic or acquired disorders of the metabolism of triglyceride-rich lipoproteins (Table 3). The defective degradation of chylomicrons with the resulting increase of triglyceride concentrations up to 10000–15000 mg/dl often causes recurrent abdominal cramps with or without acute pancreatitis, as well as hepatosplenomegaly, eruptive xanthomata, and lipemia retinalis. Laboratory tests are often disturbed by the turbid serum. The severe pain and the often necrotizing acute pancreatitis can be stopped quickly by eliminating part of the severely lipemic plasma with plasma exchange (Richter et al. 1987). Thus, the prognosis of this dyslipoproteinemia clearly depends on early diagnosis and on early treatment.

Prognosis of Atherosclerosis

The main question regarding the prognosis of dyslipoproteinemias is whether the treatment of these disorders affects the natural course of ath-

Table 3. Causes of the chylomicronemia syndrome

Inherited
  Familial lipoprotein lipase deficiency
    (type I hyperlipoproteinemia)
  Familial type V hyperlipoproteinemia
  Familial apo C II deficiency
  Familial lipoprotein lipase inhibitor

Acquired
  Diabetes mellitus
  Hypothyroidism
  Renal insufficiency
  Alcohol
  Drugs
  Dysgammaglobulinemia
  Systemic lupus erythematodes
  Lymphoma

Table 4. Lipid Research Clinic Coronary Primary Prevention Trial (LRC-CPPT). Results of treatment with cholestyramine after 7.4 years

Decrease in	%
Cholesterol	13.4
LDL cholesterol	20.3
CHD risk	19.0
Death due to coronary disease	24.0
Myocardial infarction	19.0
Positive electrocardiogram	25.0
Angina pectoris	20.0
Coronary bypass	21.0

erosclerosis. Until 1984 all the answers to this question were casuistic or indirect. At that time the data of the Lipid Research Clinic Coronary Primary Prevention Trial (LRC-CPPT) were published, proving that progression of atherosclerosis can be delayed by effective lipid-lowering drug treatment (Table 4; Lipid Research Clinics Program 1984a, b). The decrease in the risk of coronary heart disease was proportional to the reduction of serum cholesterol concentrations at ratio of 2:1 (2% reduction of CHD is achieved by 1% reduction of total cholesterol concentration). These effects have been shown to be clearly dose related, meaning that the patients taking the highest dosage of the drug (20–24 g cholestyramine) had a 19% reduction of total cholesterol and a 39.3% reduction of coronary risk.

The Helsinki Heart Study (Frick et al. 1987) is another primary prevention trial which was performed using the lipid-lowering drug gemfibrozil (which belongs to the group of fibrates). This 5-year trial demonstrated that even a moderate modification of the atherogenic lipoprotein constellation caused an impressive reduction in the incidence of coronary heart disease (Table 5).

The number of deaths definitely resulting from cardiac disease continuously decreased in the treatment-group, while they continuously increased in the placebo group. The same has been shown for the long-term follow-up of the Coronary Drug Project. Throughout the 16-year follow-up a continuously better survival curve for the niacin-treated group as compared with the placebo group became evident (Canner et al. 1986).

The Stockholm Coronary Heart Disease (CHD) Study also demonstrated a significantly lower rate of deaths from cardiovascular disease and

Table 5. The Helsinki heart study

	Change (%)
Total cholesterol	− 7.4
LDL cholesterol	− 8.2
Serum triglycerides	−34.5
Definite coronary heart disease	−33.0
Non-fatal myocardial infarctions	−37.0
Definite coronary death	−26.0
HDL cholesterol	+ 9.0

−, decrease; +, increase.

Table 6. Deaths due to acute myocardial infarction 1986 in the Federal Republic of Germany

Male	46487	14.1%
Female	33753	9.1%
Total	80240	11.4%

reinfarction rate in the treatment group as compared with the placebo group throughout the 5 years of observation (Rosenhamer and Carlson 1980).

The question remains, however, whether there is also direct evidence that lipid-lowering treatment positively affects the progression of atherosclerosis. It is clear that only small numbers of patients can be investigated by invasive diagnostic procedures such as coronary angiography. Arntzenius and coworkers (1985) did demonstrate in 39 patients that dietary treatment for 2 years inhibited progression of angiographically documented atherosclerotic lesions in those patients who had an effective lipid-lowering treatment and whose ratio of total cholesterol to HDL cholesterol was lower than 6.9.

Another angiographically controlled, secondary prevention study with cholestyramine demonstrated that in the treatment group only 12% of the > 50% stenosis showed progression as compared with 33% in the placebo group (Brensike et al. 1984).

In the CLAS study, Blankenhorn and coworkers (1987) demonstrated that after 2 years of effective treatment with diet, colestipol, and nicotinic acid the total cholesterol concentrations were lowered by 26% and LDL cholesterol by 43%, while the HDL cholesterol concentrations simultaneously increased by 37%. This was accompanied by only 10% new lesions in the drug treatment group as compared with 22% new lesions in the placebo group for coronary arteries and 18% or 30% and for venous bypasses.

In conclusion, it may be suggested that through effective lipid-lowering treatment the natural course of atherosclerosis can be retarded and possibly regression can be achieved. There is substantial evidence for the hope that also in the Federal Republic of Germany a decrease of deaths due to acute myocardial infarction (Table 6) can be achieved with an effective

Table 7. Goals of treatment (policy statement of the European Atherosclerosis Group)

		No other marked risk factors	Other marked or multiple risk factors
Total cholesterol	mg/dl	200–215	200
	mmol/l	5.2–5.7	5.2
LDL cholesterol	mg/dl	155	135
	mmol/l	4	3.5
Triglycerides	mg/dl	200	200
	mmol/l	2.3	2.3

lipid-lowering treatment and by eliminating other cardiovascular risk factors. However, the treatment goals must indeed be reached (Table 7).

References

Arntzenius AC, Kromhout D, Barth JD, Reiber JHC, Bruschke AVG, Buis B, Van Gent CM, Kempen-Voogd N, Strikwerda S, Van der Velde EA (1985) Diet, lipoproteins, and the progression of coronary atherosclerosis. N Engl J Med 312:805–811

Bilheimer DW, Goldstein JL, Grundy SM, Starzl TE, Brown MS (1984) Liver transplantation to provide low-density-lipoprotein receptors and lower plasma cholesterol in a child with homozygous familial hypercholesterolemia. N Engl J Med 311:1658–1664

Blankenhorn DH, Nessim SA, Johnson RL, Sanmaro ME, Azen SP, Cashin-Hemphill L (1987) Beneficial effects of combined colestipol-niacin therapy on coronary atherosclerosis and coronary venous bypass grafts. JAMA 257:3233–3240

Brensike JF, Levy RI, Kelsey SF, Passamani ER, Richardson JM, Loh IK, Stone NJ, Aldrich RF, Battaglini JW, Moriarty DJ, Fisher MR, Friedman L, Friedewald W, Detre KM, Epstein SE (1984) Effects of therapy with cholestyramine on progression of coronary arteriosclerosis: results of the NHLBI type II coronary intervention study. Circulation 69:313–324

Canner PL, Berge KG, Wenger NE, Stamler J, Friedman L, Prineas RJ, Friedewald W (1986) Fifteen year mortality in coronary drug project patients: long-term benefit with niacin. J Am Coll Cardiol 8:1245–1255

Frick MH, Elo O, Haapa K, Heinonen OP, Heinsalmi P, Helo P, Huttunen JK, Kaitaniemi P, Koskinen P, Manninen V, Mäenpää H, Mälkönen M, Mänttäri M, Norola S, Pasternack A, Pikkarainen J, Romo M, Sjöblom T, Nikkilä EA (1987) Helsinki heart study: primary-prevention trial with gemfibrozil in middle-aged men with dyslipidemia. N Engl J Med 317:1237–1245

Janus ED, Grant S, Lintott CJ, Wardell MR (1985) Apolipoprotein E phenotypes in hyperlipidaemic patients and their implications for treatment. Atherosclerosis 57:249–266

Kannel WB, Castelli WP, Gordon T (1979) Cholesterol in the prediction of atherosclerotic disease. New perspective based on Framingham study. Ann Intern Med 90:85

Lipid Research Clinics Program (1984a) The lipid research clinics coronary primary prevention trial results. I. Reduction in incidence of coronary heart disease. JAMA 251:351–364

Lipid Research Clinics Program (1984b) The lipid research clinics coronary primary prevention trial results. II. The relationship of reduction in incidence of coronary heart disease to cholesterol lowering. JAMA 251:365–374

McGee D (1973) The Framingham study: an epidemiological investigation of cardiovascular disease. Section 27. Department of Health, Education and Welfare publication no. (NIH) 74–618

Richter WO, Brehm G, Schwandt P (1987) Type V hyperlipoproteinemia and plasmapheresis. Ann Intern Med 106:779

Rosenhamer G, Carlson LA (1980) Effect of combined clofibrate-nicotinic acid treatment in ischemic heart disease. Atherosclerosis 37:129–138

Starzl TE, Bilheimer DW, Bahnson HT, Shaw BW jr, Hardesty RL, Griffith BP, Iwatsuki S, Zitelli BJ, Gartner JC jr, Malatack JJ, Urbach AH (1984) Heart-liver transplantation in a patient with familial hypercholesterolemia. Lancet II:1382–1383

Study Group, European Atherosclerosis Society (1988) The recognition and management of hyperlipidaemia in adults: a policy statement of the European Atherosclerosis Society. Eur Heart J 9:571–600

Acute and Chronic Effects of Diet and Physical Exercise on Plasma Lipids *

G. Schlierf and G. Schuler

Diet and physical exercise strongly influence plasma lipids and lipoproteins. In a majority of adult women and men blood cholesterol levels deviate from norm (according to recent recommendations this is 200 mg/dl) as a result of sedentary life style and faulty nutrition. In these "patients"

* Supported by Bundesminister für Forschung und Technologie, Bonn and Verein zur Förderung der Herzinfarktforschung, Heidelberg.

Table 1. Metabolic variables (secondary variables). (From Schuler et al. 1988)

Variable	Baseline	3 months	6 months	9 months	12 months
Intervention group ($n=18$)					
BMI (kg/m^2)	26.1± 2.8	24.6± 2.8	24.5± 3.0	24.5± 3.0	24.4± 3.2
Cholesterol (mg/dl)	242 ±32	201 ±38	192 ±41	199 ±31	202 ±31
LDL (mg/dl)	147 ±37	131 ±32	125 ±34	135 ±25	130 ±30
VLDL (mg/dl)	37 ±46	22 ± 9	19 ±14	18 ±11	22 ±15
HDL (mg/dl)	39 ± 6	40 ± 8	42 ± 7	42 ± 8	40 ± 7
TG (mg/dl)[a]	151 (87–303)	114 (64–250)	84 (53–200)	99 (57–268)	105 (69–304)
CHOL/HDL	6.4± 1.3	5.3± 1.4	4.7± 1.2	4.9± 0.9	5.1± 1.0
Control group ($n=17$)					
BMI (kg/m^2)	26.0± 2.5	26.5± 2.8	26.4± 2.8	26.3± 2.7	25.9± 2.4
Cholesterol (mg/dl)	241 ±30	240 ±33	243 ±33	244 ±52	258 ±53
LDL (mg/dl)	150 ±41	162 ±36	165 ±37	168 ±42	172 ±35
VLDL (mg/dl)	39 ±28	31 ±21	29 ±14	30 ±21	36 ±54
HDL (mg/dl)	36 ± 7	36 ± 7	39 ±10	36 ± 9	37 ± 9
TG (mg/dl)[a]	175 (93–350)	151 (85–342)	132 (82–277)	145 (80–650)	157 (81–2 202)
CHOL/HDL	7.0± 1.5	6.8± 1.5	6.6± 2.0	7.3± 2.5	7.5± 3.1

BMI, body mass index (weight/height^2); CHOL, total cholesterol.
[a] Median (range).

Table 2. Metabolic variables: statistical evaluation. (From Schuler et al. 1988)

Variable	Baseline	Average (6–12 months)	P vs baseline	P vs control
Intervention group (n=18)				
Body mass index (kg/m²)	26.1 ± 2.8	24.5 ± 3.0	<0.0002	<0.05
Cholesterol (mg/dl)	242 ± 32	198 ± 27	<0.0002	<0.0002
LDL (mg/dl)	147 ± 37	130 ± 28	<0.05	<0.002
VLDL (mg/dl)	37 ± 46	19 ± 12	<0.05	<0.05
HDL (mg/dl)	38.7 ± 6.3	41.1 ± 6.8	NS	NS
TGs (mg/dl)[a]	151 (87–303)	117 (71–304)	<0.005	<0.01
CHOL/HDL	6.4 ± 1.3	4.9 ± 0.9	<0.002	<0.002
Control group (n=17)				
Body mass index (kg/m²)	26.0 ± 2.5	26.2 ± 2.6	NS	
Cholesterol (mg/dl)	243 ± 30	249 ± 41	NS	
LDL (mg/dl)	150 ± 42	169 ± 34	<0.01	
VLDL (mg/dl)	39 ± 28	32 ± 29	<0.05	
HDL (mg/dl)	35.7 ± 6.8	37.2 ± 8.0	NS	
TGs (mg/dl)[a]	175 (93–350)	148 (95–1043)	<0.05	
CHOL/HDL	7.0 ± 1.5	7.2 ± 2.1	NS	

Average (6–12 months), average total of observations at 6, 9, and 12 months; CHOL, total cholesterol.
[a] Median (range).

a sensible diet, e.g., one following the recommendations of the German Society for Nutrition, combined with more physical exercise can normalize atherogenic lipoproteins or prevent their development. The following summary, which is mostly based on results obtained in studies carried out in our institution, outlines the effects and thus calls to mind the potential of a sensible way of life.

In one study we examined the effects of a low-fat diet and intensive physical exercise in 18 patients with coronary heart disease as diagnosed by angiographic examination and compared them with a control group of patients in normal care. After 12 months the mean blood cholesterol value of 242 + 32 mg/dl had dropped to the "ideal" value of 202 + 31 mg/dl. Low-density lipoprotein (LDL) levels fell to 130 mg/dl and triglycerides to 105 mg/dl. High-density lipoprotein (HDL) levels did not increase significantly, probably due to the opposing effects of physical exercise (increase) and low-fat diet (decrease). All risk parameters remained the same in the control group (Tables 1, 2). Measurement of cardiological parameters in the group under study revealed a significant increase of 21% in physical work capacity and a 54% decrease in load-induced myocardial ischemia.

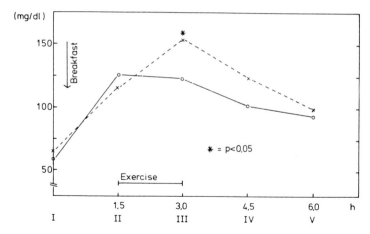

Fig. 1. Postprandial triglycerides (medians) with (*solid line*) and without (*broken line*) exercise (corrected for free glycerol). (From Schlierf et al. 1987)

The values of the control group did not change. It remains to be seen whether the observed effects can be maintained and which mechanisms are responsible for the improvement in the cardiological parameters.

A second study was undertaken to examine the acute effects of physical activity on alimentary lipemia, i.e., the postprandial increase of plasma triglycerides. In 12 healthy young men a 90-min training session (bicycle ergometry) caused a significant reduction of 34% in alimentary lipemia after a normal breakfast (Fig. 1). The determining factor is most likely the acute increase (by 42%) in lipoprotein lipase activity caused by the physical exercise as compared with the resting control group.

Since not only lipoprotein levels measured on an empty stomach, but also those measured after eating have been associated with atherogenesis, the effects described above may contribute to a long-term reduction of risk of atherosclerosis. It seems advisable when considering the German saying „nach dem Essen sollst du ruh'n oder tausend Schritte tun" (after eating one should rest or take a thousand steps) to choose the latter under these circumstances.

In summary, a sensible combination of diet and physical exercise is an appropriate means of normalizing moderately high blood fat levels (as usually seen in patients with coronary heart disease) and pathologic lipoprotein patterns. Just as life style factors lead to decompensation of lipid metabolism in the majority of patients with hyperlipoproteinemia, they may normalize metabolic disorders without resorting to drugs in the group of patients discussed here. The use of drugs can then be restricted to patients with severe disorders of fat metabolism.

References

Schlierf G, Dinsenbacher A, Kather H et al. (1987) Mitigation of alimentary lipemia by post-prandial exercise – phenomena and mechanisms. Metabolism 36:762–730

Schuler G, Schlierf G, Wirth A et al. (1988) Low-fat diet and regular, supervised physical exercise in patients with symptomatic coronary artery disease: reduction of stress-induced myocardial ischemia. Circulation 77:172–181

Present Concepts for Drug Treatment of Hyperlipidaemia in Adults

H. Greten

Hyperlipidaemias or hyperlipoproteinaemias are metabolic disorders that result from accelerated synthesis or retarded catabolism of lipoproteins which transport cholesterol and triglycerides in plasma. Elevated plasma lipoprotein concentrations deserve particular attention by the clinician as they may cause two life-threatening conditions: accelerated atherosclerosis and acute haemorrhagic pancreatitis.

Most hyperlipidaemias are primary, in the sense that the cause of the lipoprotein abnormality lies in genetic and environmental defects of synthesis or degradation of a given lipoprotein particle. According to Goldstein and Brown, primary hyperlipidaemias can be divided into two classes: (a) single gene disorders and (b) multifactorial disorders.

Secondary hyperlipidaemias are less common. A variety of underlying disorders produce secondary hyperlipoproteinaemias, the most frequently encountered forms accompanying diabetes, hypothyroidism, obesity, liver disease, alcoholism and renal dysfunction. Some 20% of patients attending the Department of Medicine, University Hospital Eppendorf, Hamburg, have elevated lipid levels due to a recognised underlying cause. In general, these lipid abnormalities improve or disappear when the underlying disorder is successfully treated. However, in some patients, especially in those with non-insulin-dependent diabetes mellitus, hyperlipidaemia may persist even after correction of hyperglycaemia. Table I summarises common clinical disorders associated with secondary hyperlipoproteinaemia.

It is now generally accepted in all countries that there is a need for strategies to reduce the rate of coronary heart disease. In principle there exist both strategies for the individual and for the population.

This report summarises possibilities for drug management of elevated blood lipids. As the risk of coronary heart disease increases greatly when more than one risk factor is present, particular attention has to be given to those individuals with other marked or multiple risk factors as regards

Table 1. Changes in lipoproteins in clinical disorders associated with secondary hyperlipo-proteinaemias

	Chylomicrons	VLDL	IDL	LDL	HDL
Diabetes mellitus	↑	↑			
Renal disease		↑	↑	↑	
Liver disease		↑	↑	↑↓	↓
Alcoholism	↑	↑			
Hypothyroidism			↑	↑	
Monoclonal gammopathies	↑	↑	↑		
Cushing disease		↑		↑	
Anorexia nervosa				↑	
Acute intermittent porphyria				↑	

VLDL, very low-density lipoprotein; IDL, intermediate-density lipoprotein; LDL, low-density lipoprotein; HDL, high-density lipoprotein.

the decision of life-long drug treatment. Several consensus documents have been published recently on the prevention of coronary heart disease. The NIH policy statement [2] defined Americans over the age of 40 with cholesterol levels between 240 and 260 mg% as being at "moderate risk" and those with cholesterol levels over 260 mg% as being at "high risk". It

Table 2. Summary of guidelines to therapy of hyperlipidaemia 1988

Goal of treatment

No other marked risk factors	*Other marked or multiple risk factors*
Cholesterol <200 mg/dl	Cholesterol <200 mg/dl
LDL-cholesterol <155 mg/dl	**LDL-cholesterol <135 mg/dl**
Triglyceride <200 mg/dl	Triglyceride <200 mg/dl

Group A: Cholesterol 200–250 mg/dl; Triglyceride <200 mg/dl

Diet	Diet
Weight control	Weight control
Drugs rarely indicated	Reassess in 3 months
Follow-up optional	If treatment goal not achieved reinforce dietary advice
	Consider use of drugs in high risk individuals

No other marked risk factors	*Other marked or multiple risk factors*

Group B: Cholesterol 250–300 mg/dl; Triglyceride <200 mg/dl

Diet	Diet
Weight control	Weight control
Reassess in 3 months	Reassess in 3 months
If treatment goal not achieved reinforce dietary advice	If treatment goal not achieved reinforce dietary advice
Consider use of drugs	Drugs often required

Table 2 (continued)

No other marked risk factors	*Other marked or multiple risk factors*
Group C: Cholesterol < 200 mg/dl; Triglyceride 200–500 mg/dl	
Correct underlying causes	Correct underlying causes
Diet	Diet
Weight control	Weight control
Reassess in 6 months	Reassess in 3 months
If treatment goal achieved reassess in 2 years	If treatment goal achieved reassess in 1 year
If treatment goal not achieved reinforce diet	If treatment goal not achieved reinforce diet
	Use of drugs in addition to dietary therapy may be **considered**
No other marked risk factors	*Other marked or multiple risk factors*
Group D: Cholesterol 200–300 mg/dl; Triglyceride 200–500 mg/dl	
Correct underlying causes	
Diet	
Weight control	
Reassess in 3 months	
If treatment goal despite dietary compliance not achieved, **consider use of drugs**	If treatment goal despite dietary compliance not achieved, **implement drug therapy**
No other marked risk factors	*Other marked or multiple risk factors*
Group E: Cholesterol > 300 mg/dl; Triglyceride > 500 mg/dl	
Correct underlying causes	
Diet	
Weight control	
Reassess in 3 months	
If treatment goal despite dietary compliance not achieved **implement** drug therapy or consider **referral** to specialized centre	

was recommended that these people be treated intensively by dietary means, and if response is inadequate, appropriate drugs should be added. The European Atherosclerosis Society (EAS) recently published a similar policy statement on the recognition and management of hyperlipidaemia in adults [3]. In this statement five treatment groups were defined. Table 2 contains the essential guidelines of these recommendations. It is important to realise that though there is no clear cut biological threshold for the begin of intervention, the EAS suggested that at levels in excess of 200 mg% intervention should begin, particularly if other marked or multiple risk factors are present.

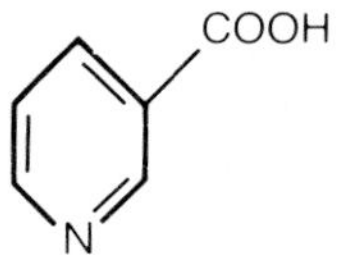

Dosis: 3-6 g

Mechanism: Lipolysis ↓, VLDL synthesis ↓

Side effects: Flush, gastrointest. symptoms

Indication: Hypercholesterolemia, hypertriglyceridemia

Efficacy: LDL-Chol. 15% ↓, VLDL 50% ↓

Fig. 1. Drug therapy for hyperlipemia using nicotinic acid (Niconacid)

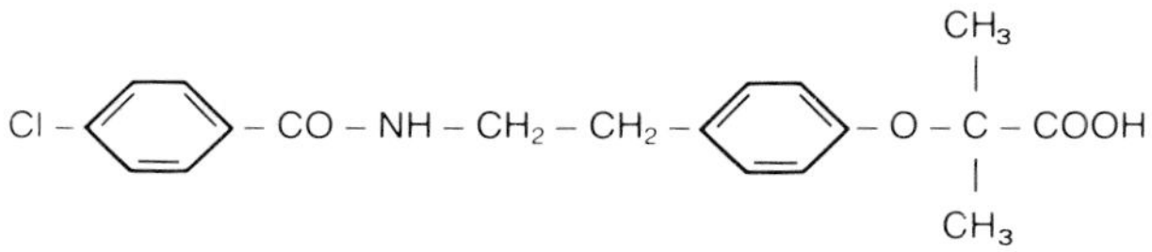

Dosis: 600 mg

Mechanism: LDL Receptor ↑, lipoprotein catabolism ↑

Side effects: gastrointest. symptoms, myositis, drug interference (anticoagulants)

Indication: Hypertriglyceridemia, Hypercholesterolemia

Efficacy: TG   50%, LDL-Chol. 12-20% ↓

Fig. 2. Drug therapy for hyperlipemia using bezafibrate (Cedur)

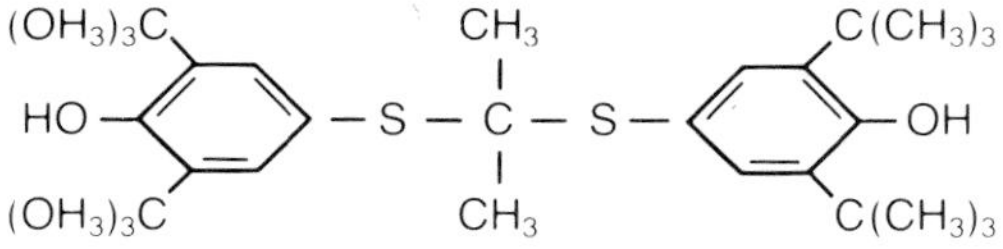

Dosis: 1 g

Mechanism: Absorption and synthesis of cholesterol ↓

Side effects: gastrointest. discomfort, ECG changes

Indication: Hypercholesterolemia

Efficacy: LDL-Chol. 10-20% ↓, HDL-Chol. ↓

Fig. 3. Drug therapy for hyperlipemia using probucol (Lurselle)

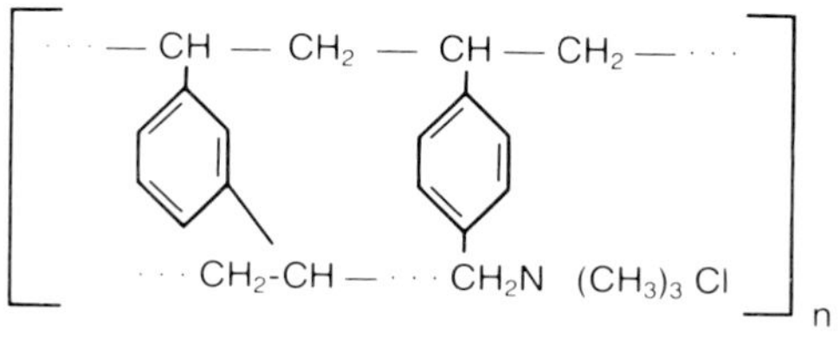

$$\left[\cdots -CH-CH_2-CH-CH_2-\cdots \atop \cdots CH_2\text{-}CH-\cdots CH_2N \; (CH_3)_3 \; Cl \right]_n$$

Dosis: 16-32 g

Mechanism: LDL Receptor ↑, blockage of bile acid recycling

Side effects: gastrointest. symptoms, drug interference

Indication: Hypercholesterolemia

Efficacy: LDL-Chol. > 20% ↓

Fig. 4. Drug therapy for hyperlipemia using colestyramine (Quantalan)

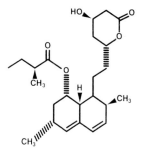

Dosis: 20-80 mg
Mechanism: β-HMG-CoA-Reductase Inhibition
                → Cholesterol Synthesis
                → LDL receptors
Side effects: gastroint. symtoms, myositis, elevation of
                transaminases (reversible)
Indication: Hypercholesterolemia
Efficacy: LDL-Chol. reduction ~ 40%

Fig. 5. Drug therapy for hyperlipemia using lovastatin (Mevinacor)

With regard to the use of lipid-lowering drugs both efficacy and safety of the particular drug should be closely monitored. A short and comprehensive summary for common drugs used in West Germany is given in Figs. 1–6.

In summary, the risk of atherosclerosis is without doubt related to the duration and degree of hyperlipidaemia. General levels for treatment have been given in consensus documents from different sources. Modern, ef-

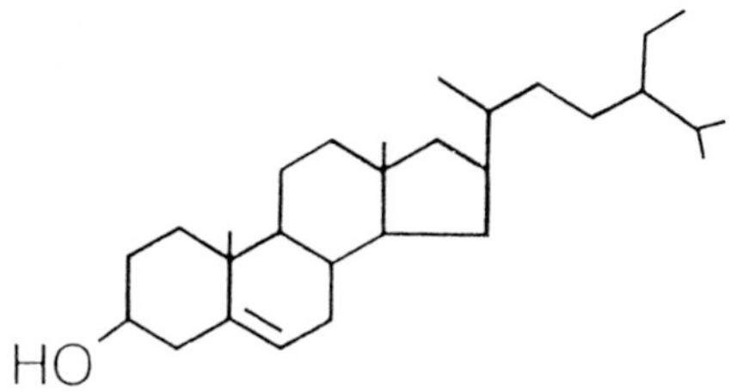

Dosis:	6 g
Mechanism:	Inhibition of intestinal cholesterol uptake
Side effects:	rare
Indication:	mild hypercholesterolemia
Efficacy:	LDL-Chol. 10-13%

Fig. 6. Drug therapy for hyperlipemia using β-sitosterol (Sito-Lande)

ficient drugs have become available for those individuals who do not respond satisfactorily to dietary regimens and changes in life style. It will be interesting to follow the development of physicians' attitudes to a more aggressive therapy for elevated lipid levels. It is the author's belief that general practitioners, internists and even cardiologists will lower their thresholds of therapeutic intervention. Hopefully this will then also lead to a reduction of coronary heart disease.

References

1. Goldstein J, Brown M (1987) In: Harrison's principles of internal medicine. 11th edn, McGraw Hill
2. NIH Consensus Conference (1985) Lowering blood cholesterol to prevent heart disease. JAMA 235:2080–2086
3. European Atherosclerosis Society (1988) The recognition and management of hyperlipidaemia in adults: a policy statement of the European Atherosclerosis Society. European Heart J 9:571–600

Comparison of Different Forms of Plasmapheresis *

C. Keller and G. Wolfram

Introduction

In 1967 plasma exchange was used for the first time to treat familial hyper-cholesterolemia (FH) by de Gennes and coworkers [2]. In 1975, 8 years later, results of plasma exchange treatment for FH over a longer period of time was reported by Thompson et al. [10] which suggested that regular elimination of LDL cholesterol from the circulation might be beneficial for homozygous FH patients. This expectation was confirmed by Thompson's observation of prolonged survival of treated homozygous patients in comparison with their untreated siblings [12].

Table 1. Plasmapheresis: Nonselective and selective elimination of LDL from plasma

Nonselective removal	
Plasma exchange	Replacement: human albumin
Selective removal	
Immune absorption	Antiapolipoprotein B antibody
Precipitation	Heparin at acidic pH
	Dextran sulfate cellulose
Filtration	Cascade
	Autologous plasma filtrate

* The studies have been supported by grants from the Deutsche Forschungsgemeinschaft and from the Bundesministerium für Forschung und Technologie (BMFT).

The regular, unspecific removal of plasma to eliminate low density lipoprotein (LDL) cholesterol stimulated the search for specific means of eradicating the pathogenic LDL. There are several methods available at the present time which are listed in Table 1 [1, 9].

At the Medizinische Poliklinik München we started plasmapheresis in 1976. Experiences with different systems will be reported.

Patients and Methods

We treated six homozygous patients, three males and three females, 8 to 44 years old, and nine heterozygous patients, 16 to 63 years old, all of which exhibited the LDL receptor defect confirmed in tissue cultures of skin fibroblasts [3]. All were suffering from severe coronary atherosclerosis. This and the fact that serum cholesterol could not be normalized with conventional drug treatment were the indications to treat heterozygous patients with plasmapheresis as well. The duration of treatment and the methods applied are apparent from the data in Fig. 1. The longest treatment in a single patient amounted to nearly 9 years.

The methods applied over a longer period of time in our department are plasma exchange with human albumin solution, filtration of the patient's plasma and subsequent use of this LDL-free plasma filtrate as a substitution solution, immune absorption technique, and heparin precipitation. Dextran sulfate cellulose columns were only used recently. These data will be reported later.

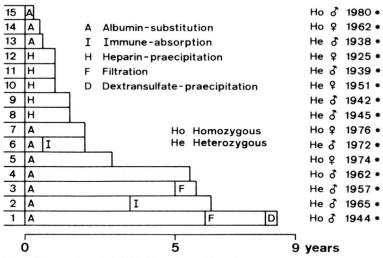

Fig. 1. Plasmapheresis in FH: 15 patients, 39 patient-years

For every treatment the anticoagulant heparin is administered with a bolus injection and the patient continuously receives heparin thereafter. During plasma exchange and with immune absorption, citric acid – dextrose A (ACD-A) solution is used in addition to avoid complement activation and platelet aggregation. The duration of a single treatment lasts up to 3 h, depending on the method used and the volume of plasma treated. Plasma volumes treated range from 2500 to 4500 ml.

Results

Figure 2 documents that the acute lowering of total and LDL cholesterol amounts to about the same percentage, regardless of the method chosen to remove. The rebound of lipids and lipoproteins following the treatment occurs at the same speed, independent of the kind of plasmapheresis employed. Opposed to Thompson's data [11], we did not find that pyridylcarbinol and cholestyramine slowed the post-exchange rebound down [4]. In the patients studied, HDL cholesterol did not increase significantly with immune absorption, as was reported by Parker et al. [7]. Measurements of

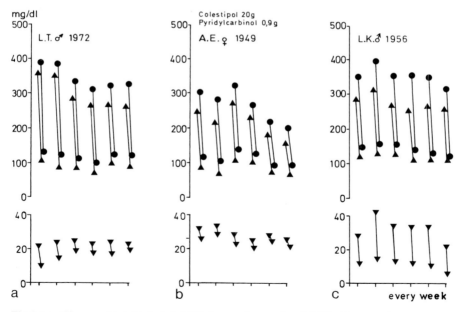

Fig. 2 a–c. Change of total (*circles*), LDL (*upward arrows*) and HDL (*downward arrows*) cholesterol with different forms of plasmapheresis: **a** immune absorption; **b** heparin-induced extracorporeal LDL precipitation (HELP); **c** human albumin

apolipoprotein B (apo B) in LDL and very low density lipoprotein (VLDL) before and after apheresis by immune absorption or heparin precipitation showed that the reduction of apo B was mainly due to a decrease of LDL apo B, whereas the reduction of VLDL apo B can be neglected (Fig. 3).

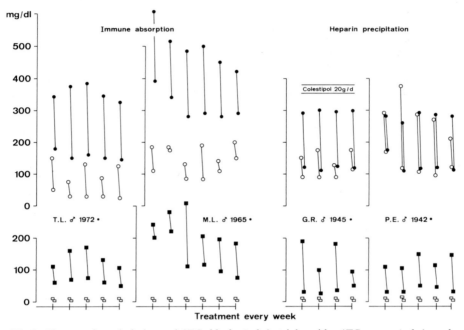

Fig. 3. Change of total cholesterol (*TC, black circles*), triglycerides (*TG, open circles*), and apolipoprotein B with different forms of plasmapheresis. Apo B LDL, *black squares*; Apo B VLDL, *open squares*

Table 2. Decrease in percent of initial values before treatment with different methods of plasmapheresis

	Plasma exchange	Immune absorption	Heparin precipitation	Plasma filtrate
Lp (a)	34	14	29	7
Total protein	23	25	18	29
Albumin	2	1	1	22
Globulin	57	1	0	39
Prothrombin time	–	23	53	–
Plasminogen	–	37	60	–
Fibrinogen	–	21	59	–
Partial thrombo-plastin time	–	0	47	–

There are significant differences with respect to lipoprotein (a) [Lp (a)], depending on the plasmapheresis method applied (Table 2). The percentage decreases of total protein, albumin, and globulin with respect to plasma exchange, immune absorption, heparin precipitation, or filtration are also listed in Table 2. The decrease of total protein occurring with immune absorption or heparin precipitation, albumin and globulin concentrations being nearly unchanged at the same time, is most likely an effect of dilution, because the patients gain about 1 kg of weight during treatment due to the infusion of sodium chloride solution. Continuous weekly plasma exchange with human albumin solution leads to a loss of immunoglobulins after months of treatment which is not replaced within 1 week by increased synthesis. Prothrombin time, partial thromboplastin time, fibrinogen, and plasminogen are decreased to a higher degree by the heparin precipitation method than by immune absorption as listed in Table 2.

Clinically we have observed a decrease in the thickness of the Achilles' tendon [8], a gradient loss of 20 mmHg across the aortic valve in the patient treated longest after a treatment period of 3 years every 2 weeks [5], and the disappearance of a soft carotid artery atheroma after 1 year of weekly therapy [6].

Discussion

The long-term application of various forms of plasmapheresis has proven to be just as safe in our institution as in other centers. The drastic decrease of LDL cholesterol is similar whatever system is used, but is different for Lp (a), depending on the kind of apheresis. The rebound speed of lipids and lipoproteins is not dependent on the apheresis method applied. Simple plasma exchange induces a loss of immunoglobulins, which is considerably less with apheresis. Though there is acute disturbance of the blood coagulation system by administering the anticoagulants heparin and ACD-A solution and by removing or precipitating clotting factors we have not observed bleeding disorders. Long-term application of the various LDL apheresis procedures will probably show whether the speed of regression of atherosclerosis is different according to the system used. So far only single case observations provide evidence that plasmapheresis is able to induce regression of atheroma. If the ongoing studies in heterozygous patients at different centers are able to show definite improvement of coronary atherosclerosis, it will be easier to define the role this costly treatment will play in the management of severe heterozygous FH.

References

1. Fuchs C, Windisch M, Wieland H, Armstrong VW, Rieger J, Köstering H, Scheler F, Seidel D (1982) Selective continuous extracorporeal elimination of low-density lipoproteins from plasma by heparin precipitation without cations. In: Gurland HJ, Lysaght MJ (eds) Plasma separation and fractionation. Karger, Basel, pp 272–280
2. De Gennes JL, Touraine R, Maunand B, Truffert J, Laudat PL (1967) Formes homozygotes cutanéo-tendineuses de xanthomatose hypercholesterolémique dans une observation familiale examplaire – Essai de plasmaphérèse à titre de traitement héroique. Soc Méd Hôp, Paris 118:1377–1402
3. Keller C, Harders-Spengel K, Spengel F, Wieczorek A, Wolfram G, Zöllner N (1981) Serum cholesterol levels in patients with familial hypercholesterolemia confirmed by tissue cultures. Atherosclerosis 39:51–58
4. Keller C, Hailer S, Demant T, Wolfram G, Zöllner N (1985) Effect of plasma exchange with and without concomitant drug treatment on lipids and lipoproteins in patients with familial hypercholesterolemia confirmed by tissue cultures. Atherosclerosis 57:225–234
5. Keller C, Schmitz H, Theisen K, Zöllner N (1986) Regression of valvular aortic stenosis due to homozygous familial hypercholesterolemia following plasmapheresis. Klin Wochenschr 64:338–341
6. Keller C, Spengel FA (1988) Changes of atherosclerosis of the carotid arteries due to severe familial hypercholesterolemia following long-term plasmapheresis, assessed by Duplex scan. Klin Wochenschr 66:149–152
7. Parker TS, Gordon BR, Saal SD, Rubin AL, Ahrens EH Jr (1986) HDL increase is more than LDL is lowered by LDL-pheresis. Proc Natl Acad Sci USA 83:777–781
8. Seidl O, Keller C, Berger H, Wolfram G, Zöllner N (1983) Xeroradiographic determination of Achilles' tendon thickness in familial hypercholesterolemia confirmed by tissue cultures. Atherosclerosis 46:163–172
9. Stoffel W, Demant T (1981) Selective removal of apolipoprotein B-containing serum lipoproteins from blood plasma. Proc Natl Acad Sci USA 78:611–615
10. Thompson GR, Lowenthal R, Myant NB (1975) Plasma exchange in the management of homozygous familial hypercholesterolemia. Lancet I:1208–1211
11. Thompson GR (1980) Plasma exchange for hypercholesterolemia: a therapeutic mode and investigative tool. Plasma Therapy 1:5–12
12. Thompson GR, Miller JP, Breslow JL (1985) Improved survival of patients with homozygous famililial hypercholesterolemia treated with plasma exchange. Br Med J 291:1671–1673

Extracorporeal Plasma Therapy in the Treatment of Severe Hyper-β-Lipoproteinemia: The HELP System

D. Seidel

Introduction

A large and convincing body of evidence links coronary risk with elevated plasma levels of both low density lipoprotein (LDL) cholesterol and fibrinogen. Cholesterol of atherosclerotic lesions originates mainly from cholesterol circulating in the blood which is bound to LDL. Most forms of hyper-β-lipoproteinemia result from a defect in extraction of LDL from plasma by the liver and the LDL receptor is now recognized as a crucial element in the control of cholesterol homeostasis [1]. Elevated levels of fibrinogen, a common phenomenon in hypercholesterolemia, increase blood viscosity and erythrocyte aggregation, thereby altering tissue perfusion in severe atherosclerotic disease. In addition, fibrinogen and its degradation products can influence prostaglandin metabolism of endothelial and vascular smooth muscle cells, facilitating platelet aggregation; it can also injure endothelial cells.

Treatment of familial hypercholesterolemia (FH) by diet and drug therapy alone is often ineffective. Encouraging results have, however, been obtained in the treatment of this disorder and atherosclerosis by plasma exchange, first introduced by De Gennes in 1967 [2].

In conventional plasmapheresis the patient's own plasma is replaced by donor plasma or more commonly by a plasma protein fraction containing albumin and a limited number of other plasma proteins. Complications may therefore arise due to the transmission of infectious diseases or the introduction of foreign proteins. It is thus not surprising that selective methods have been sought to remove LDL and return the LDL-free plasma to the patient. The different procedures that have been developed for this purpose can be classified as follows:

- Double membrane filtration
- Immunological procedures
- Polyanion adsorption
- Polyanion precipitation

On account of its large molecular weight LDL can be separated from other plasma proteins by filtration. This procedure, however, is not entirely specific and there is some loss of high density lipoprotein (HDL) and other plasma proteins of large molecular weight by the membrane filtration technique.

The immunological procedures utilize antibodies to apo B, usually immobilized on sepharose. By this LDL-apheresis system both apo-B-containing lipoprotein fractions, LDL and very low density lipoprotein (VLDL), are eliminated. The technique requires the preparation of large quantities of sterile immobilized antibodies and for economic reasons it is necessary to regenerate and reuse the columns. Each patient needs his own column. In order to avoid adverse immunological reactions it is also important that no antibodies leach into the patient.

Based on the affinity of LDL for polysulfated polysaccharides, heparin-agarose beads in the presence of calcium and immobilized dextran sulfate have also been used for the continuous elimination of LDL from plasma. Because of its immunostimulatory effect it is, however, very important that no dextran sulfate leaches from the column. For both adsorption procedures the capacity in relation to the volume of the column remains a technical problem.

The goals for the development of a new system were as follows:

1. To remove both LDL and fibrinogen with high efficiency
2. To use only disposable material and to avoid regeneration of any of the used elements
3. To avoid the use of compounds with immunogenic or immunostimulatory activity
4. To provide a technically safe and well standardized procedure

The system developed is based on a coprecipitation of LDL and fibrinogen at acid pH in the presence of heparin [3, 4]. It has been named HELP for heparin-mediated extracorporeal LDL: fibrinogen precipitation. In this article I will describe the system and our clinical experience to date in treating patients with FH.

The HELP Procedure

It has long been known that lipoproteins can be precipitated by polyanions at neutral pH with divalent metal ions. However, the different lipoprotein classes also demonstrate pH-dependent precipitation with heparin in the absence of divalent cations. The actual pH at which a particular lipoprotein can be precipitated depends upon several factors, including the ionic strength of the medium, the presence or absence of other plasma proteins, and the heparin concentration. Of the three major lipoprotein classes, high density (HDL), very low density (VLDL) and low density (LDL), the latter are precipitated at a higher pH range than the other two. A selective procedure could, therefore, be developed for LDL which is based on an increase of the positive charge on the LDL particles at low pH, allowing them to form a network with heparin and fibrinogen [3, 4].

In addition to LDL and fibrinogen only a limited number of other heparin-binding plasma proteins are coprecipitated by heparin at low pH (plasminogen, C3 and C4 complement) [4, 5]. Other proteins such as apo AI, apo AII, albumin, or immunglobulins that do not bind heparin are not precipitated under these conditions [3–5]. The HELP system has the advantage that the patient is not exposed to foreign proteins or compounds with the attendant immunological problems. It displays a high degree of reproducibility and an almost unlimited capacity, which guarantees a constant therapy independant of the clinic performing the treatment.

On the basis of our in vitro studies, an extracorporeal procedure for the continuous elimination of LDL and fibrinogen with heparin at acid pH was developed in collaboration with the B. Braun Company (Melsungen, FRG).

The HELP System

The major characteristics of the HELP system are illustrated in the flow sheet (Fig. 1). In the first step, plasma is obtained by filtration of whole blood through a 0.2-µ filter. This is then mixed continuously with a 0.3 *M* acetate buffer of pH 4.85 containing 100 IU heparin/ml. The flow rates of plasma and buffer are normally identical and range from 20 to 30 ml/min. Precipitation occurs in a small mixing chamber at a pH of 5.12 and the suspension is continuously recirculated through a 0.4-µ polycarbonate filter (from which an LDL- and fibrinogen-free, clear filtrate is obtained). Approximately 50% of the heparin required for the precipitation is removed with the precipitate while the remainder is adsorbed by passage of the plasma through an anion-exchange filter. At a pH of 5.12 the polyanionic heparin is still negatively charged on account of its strongly acidic sulfate groups, while plasma proteins will be positively charged at this pH and therefore not retained by this resin. After heparin adsorption the buffer-plasma mixture is subject to a bicarbonate dialysis and ultrafiltration. In this step excess fluid is removed, physiological pH is restored, and the acetate levels reduced before the plasma is mixed with the blood cells and returned to the patient.

D. Seidel

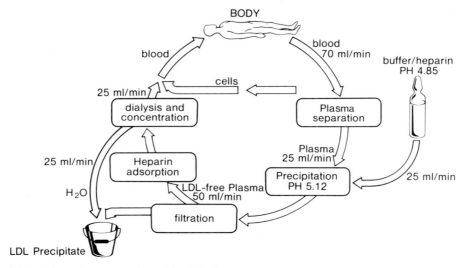

Fig. 1. Schematic presentation of the HELP system

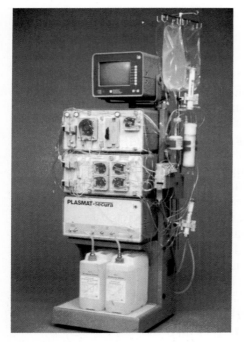

Fig. 2. Plasmat secura (Braun, Melsungen AG). Fully equipped with the disposable material

The filter and tubings required for the treatment are sterile, disposable systems that are intended for single use only, making it easy and reliable to work with and guaranteeing a standard quality for each treatment. Safety is further assured by a visual display and two microprocessors operating in parallel (Fig. 2).

The efficiency of the system for LDL and fibrinogen elimination is 100% in a single passage through the extracorporeal circuit. The total extracorporeal plasma volume of the system amounts to approximately 600 ml. In settings for the treatment of homozygous children the buffer-plasma ratio is altered to 3:1, which further reduces the plasma volume of the system. We can clear 10–15 g of cholesterol and fibrinogen in one treatment, which turns the filter striking yellow (Fig. 3). Three liters of plasma are treated in 1.5–2 h. Due to the excellent tolerance of the procedure the patients leave the hospital shortly after the end of the treatment.

Clinical Experience with the HELP System

At present five patients with FH (including one homozygous child) have been treated by HELP for up to 2 years. More than 70 patients are now undergoing treatment in 12 different centers. At present we are surveying approximately 3500 single treatments. With conventional adjuvant therapy the frequency of treatments has averaged once a week. Average pretreatment LDL cholesterol levels have been reduced by approximately 50% compared with values prior to therapy. The mean LDL cholesterol concentration to which the vascular wall is exposed is of course lower

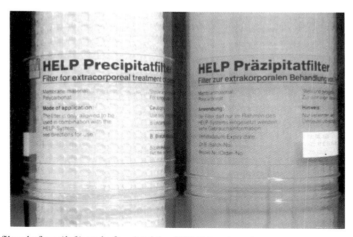

Fig. 3. Precipitation filter before (*left*) and after (*right*) a HELP treatment

Pat. St. E. ♀, 43 y. heteroz. FH.   Baseline LDL–C 473 mg/dl ± 25

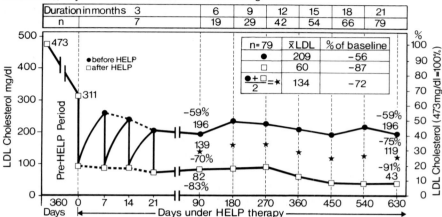

Fig. 4. Typical LDL follow-up on long-term treatment with the HELP system. The increase of LDL after the HELP treatment is almost linear within the first week. Therefore the mean values (*stars*) are calculated as LDL concentration at the end of a therapy (*squares*) plus LDL concentration before the next (*black circles*) therapy divided by two. All points given at the end of a 3-month period represent the mean of all points within this time period

Pat. Ch. J. ♀, 7 y. homoz. FH.   Baseline LDL–C 820 mg/dl ± 50

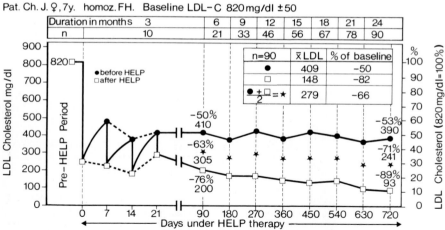

Fig. 5. LDL follow-up on long-term treatment with the HELP system in a homozygous FH child. *Black circles*, concentration before therapy; *squares*, LDL concentration after therapy; *stars*, mean LDL concentration is the concentration before therapy plus the concentration after therapy divided by 2. All points given at the end of the 3-month period represent the mean of all points within this time period

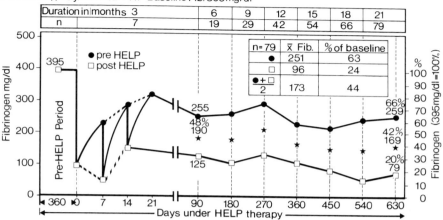

Fig. 6. Typical fibrinogen follow-up on long-term treatment with the HELP system. The increase of fibrinogen after the HELP treatment is almost linear within the first week. Therefore the mean values (*stars*) are calculated as fibrinogen concentration at the end of a therapy (*squares*) plus fibrinogen concentration before the next (*black circles*) therapy divided by 2. All points given at the end of a 3-month period represent the mean of all points within this time period

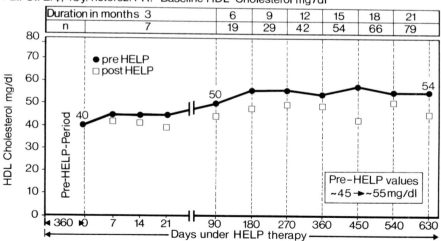

Fig. 7. HDL follow-up and long-term treatment with the HELP-system. *Black circles*, before treatment; *squares*, after treatment. All points given at the end of the 3-month period represent the mean of all points within this time period

D. Seidel

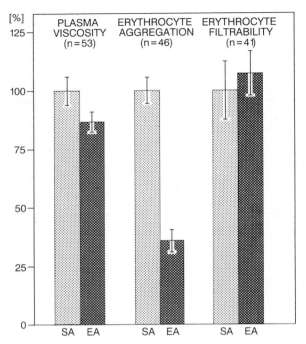

Fig. 8. Influence of the HELP treatment on plasma viscosity, erythrocyte aggregation, and erythrocyte filtrability. Values at the beginning of the therapy are expressed as 100%. *SA*, values before; *EA*, values at the end of the HELP therapy

Pat. F. E. ♂, 38 y.   Baseline LDL–C 335 mg/dl ± 15

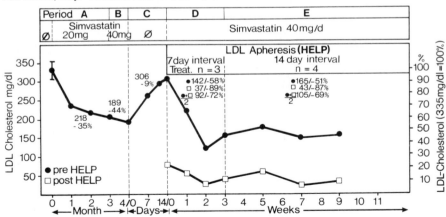

Fig. 9. Follow-up of LDL concentrations on a combined HMG-CoA reductase inhibitor/ HELP therapy of a patient with hypercholesterolemia and severe coronary heart disease. *Black circles*, before treatment; *squares*, after treatment. The approx. 70% reduction to mean LDL concentrations of 105 mg/dl from starting values 335 mg/dl is impressive

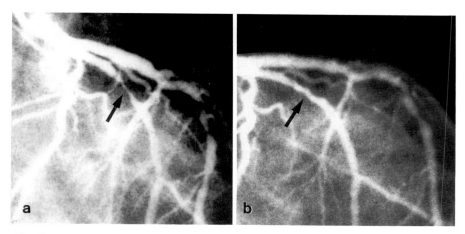

Fig. 10a, b. Coronary angiography **a** before and **b** after 2 years of HELP therapy in weekly intervals in a 30-year-old female patient. Before the treatment she had suffered from 2 coronary infarctions due to FH accompanied by severe three-vessel disease and was unable to work. She is now back at her regular job

(40% or less) since the posttreatment values may be as low as 20% or less compared with values prior to HELP therapy (Figs. 4, 5). The data for fibrinogen (Fig. 6) are very similar to the LDL data. Plasma HDL concentrations are unaffected by the HELP procedure; in some cases an increase of approximately 10%–20% may take place (Fig. 7).

The HELP treatment also significantly improves plasma viscosity, erythrocyte aggregation, and erythrocyte filtration, indicating a positive change in membrane fluidity. This may be of major clinical importance in subjects suffering from atherosclerosis [6] (Fig. 8). It is tempting to associate these data with relief from angina and improvement in exercise ECG that we have observed in most patients 2–3 months after the start of the therapy. Further studies will, however, be required to substantiate this hypothesis.

Overall treatment tolerance has been very good and no major complications have been observed after approx. 3500 treatments. Despite long-term treatment with HELP, pretreatment and levels of plasminogen C4 and C3 complement – also heparin-binding proteins – have remained stable, indicating that treatment does not lead to a deficiency of these proteins. For proteins that are not precipitated by heparin at low pH, plasma concentrations at the end of the HELP therapy were generally in the range of 80%–90% of the initial values. Samples taken 24 h after the end of the treatment showed that these proteins had retained their original level [4, 5]. Substitution of any kind has not been necessary in 2 years of experience.

Special attention has been focused on the effect of HELP on hemostasis. All posttreatment controls were typical for extracorporeal procedures. Plasma heparin levels at the end of the treatment averaged 0–0.17 IU/ml. No bleeding complications have been observed. Plasma electrolyte, hormone, vitamin, enzyme, and immunoglobulin concentrations, as well as hematological parameters remained virtually unchanged at the end of each treatment and after more than 100 weeks of treatment [4].

Combination of HMG CoA Reductase Inhibitors with HELP Therapy

With the introduction of a new generation of drugs, the HMG CoA reductase inhibitors, a new avenue of chemotherapy became available in the treatment of hypercholesterolemia [7]. As much as 30%–40% reduction of plasma and LDL cholesterol may be achieved in inherited heterozygous and in nonfamilial hypercholesterolemia [8]. Although side effects seem to be rare, in severe cases of hyper-β-lipoproteinemia, with plasma levels of 300 mg/dl LDL-C or greater, the use of such compounds may not be sufficient – in particular if the dyslipoproteinemia is accompanied by coronary sclerosis. If regression of atherosclerotic plaques is approached as a means of secondary intervention, LDL cholesterol levels of less than 140 or even less than 120 mg/dl should be reached. This is now possible with the combination of HMG CoA reductase inhibitors and an apheresis treatment with the HELP system (Fig. 9). This combination provides a new dimension for treatment of severe FH, with a reduction in mean LDL cholesterol levels of more than 70% (Fig. 9). In the combined form of therapy, intervals between each HELP treatment may be increased from 7 to 14 days depending on the synthetic rates for LDL or the severity of the disease.

Summary and Outlook

The HELP procedure provides a new means of treating severe hypercholesterolemia with the additional effect of lowering fibrinogen. It utilizes only disposable material; it retains a high degree of specificity with a 100% efficiency for LDL and fibrinogen extraction from plasma. It has the advantage that the patient is not exposed to foreign proteins or compounds with the attendant immunological problems. It displays a high degree of reproducibility and an almost unlimited capacity, guaranteeing a constant therapy independent of the clinic performing the treatment.

The potential of this cholesterol- and fibrinogen-lowering therapy will depend upon its ability to cause regression of atherosclerotic plaques. The first coronary reangiography after a 2-year HELP treatment period in a 30-year-old female patient gives support to this hope (Fig. 10). Also, sporadic observations along these lines have been reported in the literature. Nevertheless, there are as yet no large-scale controlled studies of this problem and, therefore, further work must involve multicenter, secondary intervention studies to ascertain the therapeutic effectiveness of this therapy. This is currently under investigation in a prospective nine-center study in which treatment efficacy with HELP is being checked by a well-standardized coronary angiography on 45 patients treated over a period of 2 years. Studies such as this should allow us to test whether regression of atherosclerosis can be brought about in severe hypercholesterolemia and, if so, at what LDL and fibrinogen plasma levels. We trust that the clinical benefit of the HELP system will be substantial for those patients who have problems in clearing LDL from their plasma pool and who are at the same time sensitive to high LDL levels and thus may develop premature atherosclerosis.

Acknowledgements. The development of the HELP procedure could only be accomplished by the involvement of a large number of persons from various disciplines. I am particular greatful to Drs. H. Wieland, V. W. Armstrong, T. Eisenhauer, P. Schuff-Werner, J. Thiery, K. Nebendahl, and G. Janning, University of Göttingen, and F. von der Haar, G. Roßkopf, D. Rath, W. Feller, and H. Fritze, Braun Company, Melsungen, Federal Republic of Germany.

References

1. Goldstein JL, Brown MS (1983) Familial hypercholesterolemia. In: Stanbury JB, Wyngaarden JB, Frederickson DS, Goldstein JL, Brown MS (eds) The metabolic basis of inherited disease, 5th edn. McGraw Hill, New York
2. De Gennes J-L, Touraine R, Maunand B et al. (1967) Formes homozygotes cutaneotendineuses de xanthomatose hypercholesterolemique dans une observation familiale exemplaire. Essai de plasmapherese à titre de traitment heroique. Bull Mem Soc Hop (Paris) 118:1137–1402
3. Seidel D, Wieland H (1982) Ein neues Verfahren zur selektiven Messung und extrakorporalen Elimination von Low Density Lipoproteinen. J Clin Chem Clin Biochem 20:684–685
4. Eisenhauer T, Armstrong VW, Wieland H, Fuchs C, Scheler F, Seidel D (1987) Selective removal of low density lipoproteins (LDL) by precipitation at low pH: first clinical application of the HELP system. Klin Wochenschr 65:161–168
5. Armstrong VW, Seidel D (1987) A novel technique for the extracorporeal treatment of familial hypercholesterolemia. In: Schlierf G, Mörl H (eds) Expanding horizons in atherosclerosis research. Springer, Berlin Heidelberg New York
6. Schuff-Werner P, Schütz E, Seyde WC, Eisenhauer T, Janning G, Armstrong VW, Seidel D (1988) Improved hemorheology associated with a reduction in plasma fibrinogen and

LDL in patients being treated by heparin-induced extracorporeal LDL precipitation (HELP). Eur J Clin Invest (Submitted)
7. Endo A, Kuroda M, Tsujita Y (1976) ML-236A, ML-236B and ML-236C, new inhibitors of cholesterogenesis produced by penicillium citrinum. J Antibiot 29:1346–1348
8. Havel RJ et al. (1987) Lovastatin (mevinolin) in the treatment of heterozygous familial hypercholesterolemia. A multicenter study. Ann of Intern Med 107:607–615

LDL Apheresis: 7 Years of Clinical Experience

H. Borberg, K. Oette, and W. Stoffel

Immune specific, on-line elimination of low-density lipoprotein (LDL) was originated by Stoffel, who between 1978 and 1981 experimentally investigated the therapeutic application of affinity chromatography [9]. This development led to its first application in patients in 1981 [10]. At that time the term LDL apheresis was introduced by the group at Köln, FRG, and since then has been reserved for immune-specific adsorption. The technical development led from a single, large apheresis column which was desorbed following therapeutic application (as were the subsequently introduced four small columns loaded sequentially) to the standardized application of two columns loaded and desorbed alternately during treatment under automated control [1]. Thus, the system became sufficiently practical to expand treatment to 10 patients at Köln during the following 5 years. The interest shown by foreign centers led to the introduction of this technique in the USA and the USSR. Starting in 1986 an additional multicenter trial was performed at five German universities, supported by the Federal Ministry for Research and Technology (Table 1).

Table 1. History of LDL apheresis (immune-specific adsorption)

Time span	Development	Reference
1977–1981	Column development	[9]
1981	Initial patient treatments	[10]
1981–1983	Technical development	[1]
1981–1986	Phase I trial	[4]
From 1986	Phase II trial	

Table 2. Technical features of LDL apheresis

Primary separation	
Plasma separation	Cubital veins (no artificial access)
Separation systems	Continuous flow centrifugation
	Flat-sheet membrane separation
Whole-blood flow rates	50–80 ml/min
Plasma flow rates	30–50 ml/min
Repetitive differential separation	
Number of columns per patient	2 columns
Number of columns per treatment	2 (1) loadings per column = 4 (3) per therapy
Life expectancy of columns	Approx. 50 treatments = 1 year
Desorption flow rates	80–100 ml/min
Capacity	Up to a maximum of 48 g LDL cholesterol per therapy

Table 3. Clinical/chemical characteristics of specific immunoadsorption (LDL apheresis)

Individual treatment
  Lowering LDL level to desired (normal–subnormal) blood levels
  No loss of normal plasma proteins or HDL

Long-term treatment
  Steady state after five to seven therapies if performed weekly
  Daily increase of LDL cholesterol
    In heterozygous patients   24 mg (19–29 mg)/day
    In homozygous patients   33 mg (27–37 mg)/day
  Average total cholesterol between two treatments 200 mg/dl
  Number of days less than 200 mg/dlLDL cholesterol: 1–5.1
  Number of days less than 100 mg/dl LDL cholesterol: 0–2.5
  In about 75% of the patients increase of HDL

As the technical aspects of LDL apheresis have already been described elsewhere [9, 2, 3], only some essential points will be mentioned here. The primary separation of the plasma is performed, essentially, with continuous-flow centrifugation or flat-sheet membrane filtration, permitting plasma flow rates of 30 to 45 ml/min in spite of the relatively low blood flow rates of 50 to 80 ml/min. It is of particular importance that the patients do not need an artificial access to the circulation and that the extracorporeal system operates with the blood flowing from one cubital vein to the other (Table 2).

The differential separation of the plasma is performed in a repetitive fashion, each patient having two columns loaded once or twice during each treatment. The columns can generally be used for about 50 treatments, which means for about 1 year, and have a capacity permitting the elimination of a maximum of 48 g LDL cholesterol per therapy. The desorption

Table 4. Safety of LDL apheresis

Exclusion of
  Complement activation
  Sensitization
  Particle leakage

Application of a completely automated adsorption–desorption device under electronic control

Sterile and pyrogen-free manufacturing, treatment and storage of columns

---

flow rate of the aqueous solution is somewhere between 80 to 100 ml/min and, thus, essentially higher than the plasma flow rates during the loading procedure (Table 3).

The safety of LDL apheresis has been sufficiently verified by laboratory examinations (Table 4) and by the number of treatments, exceeding 5000, already undertaken at Köln. The obvious discrepancy between the complement activation of the carrier material Sepharose 4bCL in vitro and the lack of side effects due to complement activation in vivo has been explained. The main portion of the complement split products which develops during the loading process binds to the column and is eliminated from the extracorporeal system during desorption, whereas the remaining amount is not clinically relevant [6–8]. A sensitization to polyclonal sheep antibody occurred in some patients, but remained insignificant, especially when a change in the production process was introduced. Particles were not washed out from the columns. The application of a completely automated, electronically controlled adsorption/desorption device with separate circuits for patient plasma and desorption fluid guarantees the safety, simplicity, and economical application of the technique. It is essential to know that the technical system is extremely versatile and can be used for other sorption systems as well. Sterile and pyrogen-free production and handling during therapy and storage eliminate corresponding problems. Because the system can be reused, it is particularly economical. As mentioned before, more than 5000 procedures performed at Köln and several thousand at other institutions demonstrate the usefulness of the technique.

The third major aspect of LDL apheresis, besides versatility of the technical system and its economical application, derives from the capacity to obtain normal and subnormal values, especially in homozygous patients, by the end of each treatment. As the process of adsorption is specific, normal plasma constituents and HDL are not lost. Changes in any clinical/chemical parameters are due to hemodilution, not to real loss or to removal by the extracorporeal system. During long-term treatment per-

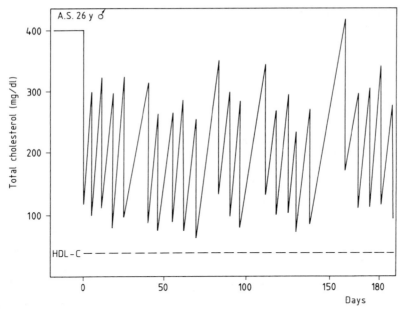

Fig. 1. Total Cholesterol under longterm LDL-Apheresis

formed without the application of drugs once per week, a steady state between elimination and synthesis rate of LDL cholesterol was established after the first five to seven treatments. This led to normal cholesterol values between treatments, depending, however, on the treatment goal for each individual person (Fig. 1). The average increase in LDL cholesterol was somewhere between 19 and 29 mg/day in heterozygous patients, but, as anticipated, greater 27 to 37 mg/day in homozygous patients. Most important for the regression of coronary heart disease is, that the LDL cholesterol levels can be kept at very low levels, for instance below 100 mg/dl for 2.5 days or 200 mg/dl LDL cholesterol for up to 5 days. In about 75% of patients an increase in HDL was observed.

Up to 1986, 10 patients were treated under the conditions mentioned above (Table 5). It should be pointed out that these treatments were not performed as a well-planned clinical trial, but were intended to demonstrate the specificity, safety, and clinical usefulness of the principle. In spite of the differences in recruitment and treatment periods some interesting results were obtained from a retrospective evaluation performed in 1986. At this time clinical symptoms had improved in more than two thirds of the patients under continuous treatment with LDL apheresis. All but one patient are now asymptomatic. The physical and functional performance of

Table 5. Specific LDL apheresis in patients with familial hypercholesterolaemia type IIa (1986)

Total patients		10
Sex	Female	7
	Male	3
Age (years)	15–57	
	$(30.3 \pm 13.9 \text{ years})$	
Homozygous		4
Heterozygous		6
Treatment period	5–54 months	
	(35.7 ± 17.4)	

the patients had improved in more than 50% of the cases. Xanthomas of the skin and the Achilles tendons had been considerably reduced or completely eliminated in all cases. Hemodynamics and left ventricular function had not deteriorated if already poor and remained normal if normal before treatment. A progression of coronary sclerosis is rare in patients undergoing treatment with LDL apheresis, occurring mainly in those patients undergoing short-term therapy so far. In 50%–70% of patients the condition remained the same, and in 30%–45% a measurable regression can be demonstrated [5]. These results will be updated during the course of the next months to include the latest developments. The sooner specific LDL apheresis is initiated, the more likely it is, that progression of coronary sclerosis can be prevented. Whereas regression of coronary heart disease was observed in younger patients only, secondary prevention is important in older patients after bypass surgery or percutaneous transluminal coronary angioplasty (PTCA). The clinical observations made in patients undergoing long-term LDL apheresis, obtained without drugs support the lipid theory of atherogenesis.

Based on the information available so far the following variables of regression of atherosclerosis in patients with familial hypercholesterolemia may be anticipated:

1. Age
2. Genetic background
3. Average cholesterol level between treatments
4. Length of treatment period
5. Clinical status at initiation of apheresis (extent of xanthomas and coronary heart disease)

As it is known that homozygous patients do not respond to drug therapy in an appropriate fashion, long-term LDL apheresis is indicated for

Table 6. Combination of LDL apheresis and drug therapy investigated drug regimens

One drug	
Fenofibrate (Lipanthyl)	3×100 mg/day
Colestyramine (Quantalan)	$3 \times\quad 8$ g/day
PMT (Ronicol retard)	3×714 mg/day
MK 803 (Mevinolin)	1.)$2 \times\quad 20$ mg/day
	2.)$2 \times\quad 30$ mg/day
Combination of two drugs	
Fenofibrate + Colestyramine	
PMT + Colestyramine	
Fenofibrate + PMT	
Colestyramine + PMT	
Combination of three drugs	
Fenofibrate + Colestyramine + PMT	

PMT, pyridolmethanoltartrate.

both receptor-negative and receptor-defective patients. The treatment should be initiated as early as possible, which means at the time when early xanthomas are evident and prior to the onset of coronary heart disease. Heterozygous patients are normally treated with diet and drug therapy. However, LDL apheresis is indicated when conventional lipid-lowering therapy with diet and drugs cannot decrease cholesterol levels to the normal range, especially in incipient coronary heart disease. If extracorporeal elimination is applied without adjuvant therapy, a consistent decrease of the average LDL cholesterol level between two treatments may not be obtained to the extent necessary for regression. Posttreatment values are below normal for a few days and return to elevated levels during the second half of the week. Thus, it appeared to be of interest to examine whether additional drug therapy could slow down the increase of LDL cholesterol so that either the length of a single apheresis could be shortened with otherwise constant treatment intervals or the interval between two therapies prolonged, thus maintaining low cholesterol levels, which appear to be necessary for the predictive regression according to the conditions set. After finishing our phase I trial, and in spite of the negative selection of these patients, none of whom had responded properly to either diet or conventional drug therapy, an additional decrease of the average LDL cholesterol of about 10%–15% was obtained (Table 6). Combinations of drugs demonstrated further reductions of up to 35% (Table 7). The modern HMG-COA-reductase inhibitors were not much better than conventional

Table 7. Combination of LDL apheresis and drug therapy: additional decrease of total cholesterol applying different regimens

	Responders n (heterozygous)	Additional decrease of total cholesterol
One drug		
Fenofibrate	4 (3)	15.5%
Colestyramin	3 (2)	21.6%
PMT	2 (2)	10.4%
Mevinolin	5 (3)	28.7%
Two drugs		
Fenofibrate + PMT	1 (1)	35.7%
Fenofibrate + Colestyramin	2 (2)	10%

PMT, pyridolmethanoltartrate.

combinations; however, the rate of side effects was considerably lower when applying Mevinolin.

In summary, we believe that due to its specificity, efficacy, economical application and clinical usefulness LDL apheresis is a convincing therapeutic approach. It not only offers patients without a therapeutical alternative a form of treatment, but also provides additional support for the lipid theory of atherosclerosis. Furthermore, it serves as a model for the specific elimination of other pathogenic substrates as does, for instance, digoxin apheresis, which was already successfully performed in our unit. Although limited capacity and uneconomical application prevented the introduction of adsorption technology into extracorporeal therapy in the past, a combination of biochemical and technical developments have brought about a fundamental change.

References

1. Borberg H (1983) The development of an automated therapeutic immunoadsorption system using the LDL-apheresis model. Eur J Clin Invest 13,2(II):A39
2. Borberg H, Stoffel W, Oette K (1983) The development of specific plasma immunoadsorption. Plasma Ther Transfus Technol 4:459–466
3. Borberg H, Gaczkowski A, Hombach V, Oette K, Stoffel W (1986) LDL-Apherese. Ärztl Lab 32:57–62
4. Borberg H, Gaczkowski A, Hombach V, Oette K, Stoffel W (1988) Treatment of familial hypercholesterolaemia by means of specific immunoadsorption. J Clin Apheresis 4:59–65
5. Hombach V, Borberg H, Gaczkowski A, Oette K, Stoffel W (1986) Regression der Koronarsklerose bei familiärer Hypercholesterinämie Typ IIa durch spezifische LDL-Apherese. Dtsch Med Wochenschr 45:1709–1715

6. Kadar JG (1988) Biocompatibility studies during specific immunoadsorption. 2nd international congress, World Apheresis Association, Ottawa, 18–20 May 1988, Abstracts, p 9
7. Kadar JG, Späth PJ, Gaczkowski A, Oette K, Borberg H (1988) Biocompatibility studies on a clinically well tolerated extracorporeal system. Plasma Ther Transfus Technol 8,4:307–318
8. Kadar JG, Späth PJ, Borberg H (In preparation)
9. Stoffel W, Demant T (1981) Selective removal of apolipoprotein B-containing serum lipoproteins from blood plasma. Proc Natl Acad Sci USA 78:611–615
10. Stoffel W, Borberg H, Greve V (1981) Application of specific extracorporeal removal of low density lipoprotein in familial hypercholesterolaemia. Lancet II:1005–1007

Disturbances of Purine Metabolism

Clinical Aspects of Gout *

H. F. WOODS

Introduction

This paper considers some clinical aspects of gout with special reference to the epidemiology of hyperuricaemia and the clinical presentation of gout and its differential diagnosis.

Epidemiology of Hyperuricaemia and Gout

Knowledge concerning the incidence of hyperuricaemia has been obtained through the screening of populations and the common use of auto-analysers. These sources of information have shown that populations contain many individuals whose serum urate concentration is high. The definition of what constitutes a high urate concentration presents some problems. Population surveys of normal healthy individuals have shown that there is a unimodal distribution of serum urate which has a positive skew in men but not in women (Mikkelsen et al. 1965; Hall et al. 1967). Within such distributions the values signifying hyperuricaemia lie within the upper tail.

To some extent the definition of uricaemia is arbitrary, being dependent upon the position assigned to the upper limit of normal. An attempt to define such a limit will have to take into consideration the factors which influence the serum urate concentration, including the sex of the subjects studied and their race, body shape, social class, diet and environment, in addition to the method of urate determination. Hyperuricaemia can be defined in three ways (Woods et al. 1986):

---

* This paper is dedicated to my friend and colleague, Nepomuk Zöllner, who has done so much to help clinicians to understand the nature of gout and how best to treat it.

1. Using statistical methods
2. In terms of the risk of developing disease which is caused by high serum urate concentrations
3. As the level of serum urate at which the risk of developing complications is greater than the risk of treatment designed to lower the serum urate

Statistical Definition of Hyperuricaemia

When considering the distribution of serum urate concentrations within a population, one definition of hyperuricaemia could be a concentration which is more than two standard deviations above the mean for that population. This definition does not hold for the skewed distribution of urate concentration in males. In defining hyperuricaemia for men, non-parametric statistical methods have been used to allow for the population skew (Woods et al. 1986; Gribsch and Zöllner 1975).

In many instances and arbitrary decision has been made and frequently quoted upper limits of normal are 7 mg dl^{-1} (0.42 mmol l^{-1}) for males and 6 mg dl^{-1} (0.36 mmol l^{-1}) for females (Wyngaarden and Kelley 1983).

When these limits are applied to population data, it is clear that hyperuricaemia is a common metabolic abnormality. A general guide for European and American populations would be that 5% of the population have hyperuricaemia using the criteria given above. There are exceptions to this in that some populations and subgroups have a higher prevalence, such as Maoris (40%; Prior et al. 1966) and hospital patients (13.2%; Paulus et al. 1970). A comprehensive list of serum urate values in populations, classes and groups has been compiled by Wyngaarden and Kelley (1978).

Hyperuricaemia and Gout

In males and females the prevalence of articular gout and that of urate stones is related to the serum urate concentration. These relationships have been demonstrated through single point and longitudinal studies. Thus Zalokar et al. (1972), using data derived from a single determination of serum urate in a large male population, showed a tenfold greater prevalence of articular gout in those with serum urate greater than 10 mg dl^{-1}, when compared with those with serum urate of 7.0 to 7.9 mg dl^{-1}.

In a longitudinal study of untreated patients, Hall et al. (1967) determined the prevalence of gouty arthritis and urate stones in the population of Framingham during a 14-year period. Figure 1 shows the results for the male subgroup and demonstrates that the risk of developing the articular

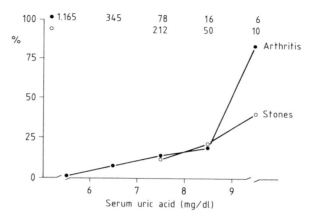

Fig. 1. The prevalence of gouty arthritis (*filled circles*) and renal stone (*open circles*) in relation to serum uric acid concentration. The data are those of Hall et al. (1967). The *figures* show the number of individuals studied at each serum uric acid concentration. (After Woods et al. 1986)

and renal manifestations of gout increases with increasing serum urate concentration. The data for females showed a similar pattern, but the risks of developing the manifestations of gout were smaller.

The relative annual risk of developing renal stones in subjects with asymptomatic hyperuricaemia and those with gout has been calculated by Fessel (1979). Fessel's results suggest that the presence of asymptomatic hyperuricaemia increases the risk of developing renal stones threefold when compared with a control group. In subjects with established gout, the risk of developing stones was more than twice that in the group with asymptomatic hyperuricaemia.

Manifestations of Gout

The manifestations of gout have been summarised by Nuki (1988) and are set out in Table 1. In this section of the paper some of the manifestations are considered in more detail.

Acute Attack of Gout

An attack of articular gout is a dramatic clinical event which usually follows a recognisable progression of symptoms and signs. An attack may be

Table 1. Major manifestations of gout. (After Nuki 1988)

1. Acute inflammatory arthritis, tenosynovitis, bursitis or cellulitis
2. Chronic, erosive, deforming arthritis associated with periarticular and subcutaneous urate deposits (tophi)
3. Nephrolithiasis and urolithiasis due to deposition of uric acid crystals from urine at acid pH
4. Chronic renal disease and hypertension

preceded by a prodromal period during which the patient may have generalised transient aches and pains in joints and stiffness of the muscles. The joint or joints later to be involved in the acute attacks can sometimes be identified because they are the site of short-lived episodes of "jabbing" or "shooting" pain. The attack itself is of rapid onset over the course of a few hours and often during the night. The affected joint becomes hot, red and tender and excruciatingly painful. The pain is probably one of the most intense experiences by humans and often appears to the sufferer to be centred over a larger area than the affected joint. The patient finds that no position or posture gives relief from the pain.

The maximum intensity of the pain occurs during the first 2 or 3 days and wanes thereafter, the attack lasting up to 10 days in all. The tissues surrounding the affected joint become oedematous and inflamed, and as the swelling subsides there may be intense itching in the tissues. Tendons adjacent to the joint may become inflamed to give a tenosynovitis.

Following an attack the affected joint may be perfectly normal in size, shape and movement, but repeated attacks and the development of periarticular erosions and tophi can lead to chronic joint disease.

The acute attack most frequently affects a single joint, usually a peripheral one. However, study of large series has shown that in about 20% of cases an attack of gout may involve more than one joint and may involve large joints such as the knee and hip. Joints such as those of the jaw and at the sternoclavicular articulation can be involved.

The differential diagnosis of the acute attack depends upon whether the diagnosis of gout has been established. For a first attack involving the foot, a diagnosis of march fracture is often made, the patient being referred to an accident department or orthopaedic clinic. When the surrounding tissues are inflamed and oedematous a diagnosis of cellulitis can be made, and in an attack involving a single joint an acute infection of that joint is part of the differential diagnosis.

In an established case of gout, the acute attack is usually correctly diagnosed (by the patient) and the main differential diagnosis centres around the chronic joint changes and occurrence of tophi. It is the latter which give rise to the main differential diagnoses of Heberden's nodes (osteoarthritis),

Table 2. Exciting (provoking) factors in acute gout

Trauma
Alcohol ingestion
Drugs
Surgery
Dietary excess
Infection
X-ray and chemotherapy
Rapid weight loss

subcutaneous calcium deposits and tendon xanthomata and rheumatoid nodules. In many instances a precipitating factor can be identified. A list of common factors is given in Table 2.

Implications of the First Attack

When faced with a patient who has suffered a first attack of articular gout the clinician has to advise as to the likelihood of that patient suffering a further attack. When the patient has hyperuricaemia and a first attack, the decision to start long-term treatment will depend upon the presence of systemic complications of gout, particularly renal disease, and the risk of that patient having further attacks of articular gout. In a longitudinal study, Gutman (1973) determined the percentage of patients having a recurrence of acute gout in relation to the length of the intercritical period (the time between acute attacks). Gutman's results show that following a first attack, 78% of patients suffer a second attack within 2 years.

X-Ray Changes in Gout

In the early stages of articular gout the X-ray changes are usually limited to soft tissue swelling which resolves with the acute attack. Recurrent attacks can be followed by permanent changes, which include periarticular erosions, the formation of tophi and secondary degenerative arthritis leading to permanent joint disease. These changes are illustrated in Fig. 2. Although the acute attack of gout and the manifestations of chronic articular gout may be mistaken for other types of arthritis, these X-ray changes are typical and an X-ray examination is essential in making a definitive diagnosis.

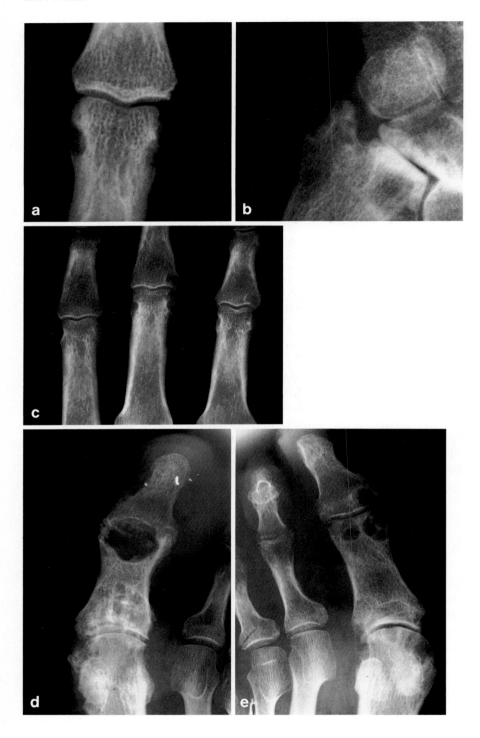

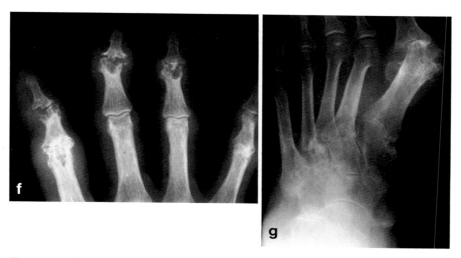

Fig. 2 a–g. X-Ray changes in gout. The plates illustrate some typical examples of the X-ray changes seen in articular gout. **a–c** Periarticular erosions; **d, e** tophi; **f, g** destructive arthritis with secondary degenerative changes

Chronic Tophaceous Gout

Repeated acute attacks may result in the development of chronic tophoaceous gout in which deposition of urate leads to a gross asymmetric distortion of the joints which classically occurs in the fingers. Tophi can occur in the helix of the ear, in periarticular tissues, in bursae and tendon sheaths and in subcutaneous tissues, as described by Zöllner (1960). There is evidence that the formation of tophi is related to the serum urate concentration. The use of drugs to lower the urate concentration has altered the natural history of gout in that the formation of tophi can be prevented and established tophi can be reduced in size. Some authorities believe that the incidence of chronic tophaceous gout is declining.

Acknowledgement. The author is grateful to Dr. Peter Ward of the Department of Radiology, The Royal Hallamshire Hospital, for his generous provision of X-ray plates.

References

Fessel WJ (1979) Renal outcomes of gout and hyperuricemia. Am J Med 67:74–82
Griebsch A, Zollner N (1973) Normalwerte der Plasmaharnsäure in Süddeutschland. Vergleich mit Bestimmungen vor zehn Jahren. Z Klin Chem Klin Biochem 11:348–356

Gutman AB (1973) The past four decades of progress in the knowledge of gout, with an assessment of the present status. Arthritis Rheum 16:431–435

Hall AP, Barry PE, Dawber TR, McNamara PM (1967) Epidemiology of gout and hyperuricemia. Am J Med 42:27–37

Mikkelsen WM, Dodge HJ, Volkenburg H (1956) The distribution of serum uric acid values in a population unselected as to gout or hyperuricemia. Am J Med 39:242–251

Nuki G (1987) Disorders of purine metabolism. In: Weatherall DJ, Ledingham JGG, Warrell DA (eds) Oxford textbook of medicine, 2nd edn. Oxford University Press, Oxford, vol 1:9 123–9 135

Paulus HE, Coutts A, Calabro JJ, Klinenberg JR (1970) Clinical significance of hyperuricemia in routinely screened hospitalized men. JAMA 211:277–281

Prior IAM, Rose BS (1966) Uric acid, gout and public health in the South Pacific. N Z Med J 65:295–300

Woods HF, Bax NDS, Jackson PR (1986) When to treat hyperuricemia. Vehr Dtsch Ges Inn Med 92:497–502

Wyngaarden JB, Kelley WN (1983) Definition and significance of hyperuricemia. In: Stanbury JB, Wyngaarden JB, Fredrickson DS, Goldstein JL, Brown MS (eds) The metabolic basis of inherited disease, 5th edn. McGraw-Hill, New York, p 1063–1064

Zaloker J, Lellouch J, Claude JR, Kuntz D (1972) Serum uric acid in 23 923 men and gout in a subsample of 4 257 men in France. J Chronic Dis 25:305–312

Zollner N (1960) Moderne Gichtprobleme: Ätiologie, Pathogenese, Klinik. In: Heilmeyer L, Schoen R, DeRudder B (eds) Ergebnisse der inneren Medizin und Kinderheilkunde, 14th edn. Springer, Berlin New York, p 376

Diet and Drug Treatment of Gout

W. Gröbner

The treatment of gout has two goals, immediate control of acute attacks of gouty arthritis and long-term control of hyperuricaemia

Treatment of Acute Gouty Arthritis

In practice colchicine, indomethacin or phenylbutazone are used for the management of acute attacks of gouty arthritis. The therapeutic effects of these drugs are fundamentally anti-inflammatory in nature and take place without detectable influence on uric acid metabolism, except in the case of phenylbutazone which is also moderately uricosuric in a high dosage. When the diagnosis is uncertain, colchicine is recommended because this drug has a high degree of specifity in gout. The maximum tolerated dose of colchicine ranges from 6 to 8 mg/day. The most frequent toxic effect is diarrhoea.

Indomethacin in a dose of 300–400 mg/day is also effective in acute gout. The dosage of phenylbutazone also must be high (up to 1 g/day) to be effective. Side effects like fluid retention and gastric discomfort must be considered.

Control of Hyperuricaemia

In the long-term treatment of hyperuricaemia and gout, serum urate should be lowered to a value of about 5–5.5 mg/100 ml. This can be achieved by decreasing the intake of dietary purines, by decreasing the synthesis of uric acid using xanthine oxidase inhibitors or by increasing the renal excretion of uric acid using uricosuric drugs.

Diet

The dietary treatment of hyperuricaemia has three goals, reduction of purine intake, weight reduction in obese patients and reduction of alcohol consumption. For the investigation of the influence of dietary purines on serum uric acid and urinary uric acid excretion isoenergetic, purine-free formula diets supplemented with purines have been used (Zöllner and Gröbner 1970; Zöllner et al. 1972; Zöllner 1975).

With an isoenergetic, purine-free formula diet, plasma uric acid and urinary uric acid excretion fall immediately. Within about 8–10 days new constant levels (plasma uric acid around 3.1 mg/100 ml, urinary uric acid excretion around 300 mg/day) are approached. When ribonucleic acid (RNA) is added to the basal diet, the plasma uric acid level rises until a new equilibrium is reached after about 1 week (Fig. 1). Concomitantly, uric acid excretion rises. If the amount of nucleic acid supplementation is increased, there is a further rise in the plasma level (Fig. 1) and excretion of uric acid. Over a range of 0–4 g daily supplementation with RNA, the rises are proportional to the amount of the nucleic acid supplemented. When desoxyribonucleic acid (DNA) is used instead of RNA for the same experiment, the rise in plasma uric acid (Fig. 1) and urinary uric acid excretion is smaller.

Mathematical analysis of these data reveals that during administration of RNA plasma uric acid rises by 0.9 mg/100 ml per 1 g RNA, while for

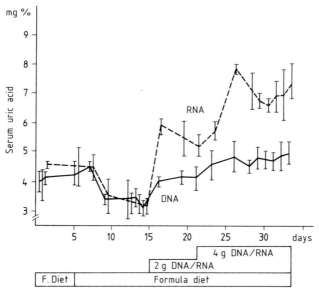

Fig. 1. Influence of dietary RNA or DNA on serum uric acid level. (From Zöllner et al. 1972)

DNA the figure is only 0.4 mg/100 ml. Uric acid excretion rises by 140 mg/day after adding 1 g RNA, while after DNA the rise is only 68 mg/day.

When the same experiments are performed with the purine ribotides adenylic or guanylic acid (AMP or GMP), the increase of serum uric acid and urinary uric acid excretion is larger than after administration of nucleic acids. The different effects of RNA, DNA and the purine nucleotides on serum uric acid and urinary uric acid excretion are probably due to different absorption rates from the gut. Compared to the purines administered, roughly 50% of those from RNA, 25% of those from DNA and 80% of those from AMP or GMP are excreted in the urine.

If hyperuricaemic persons are used for the same type of experiments, the response of plasma uric acid to RNA or DNA is much stronger, the slope of the regression line being more than 50% steeper than in normal subjects (Zöllner et al. 1972). These experimental data demonstrate that not all dietary purines have the same effect on purine metabolism. Therefore, the total purine values in conventional food tables are of restricted usefulness for dietary purposes. A food high in purines because of its DNA content may thus influence uric acid parameters less than one which is somewhat lower in purines, but which mainly contains RNA (Zöllner et al. 1972; Colling and Wolfram 1987).

In practice we recommend to our gout patients a moderate intake of purines which includes one meal of meat (100–150 g) per day. Milk and milk products, which are purine-free foods, should be the main protein source. Obesity should be corrected and alcohol consumption reduced.

In patients with asymptomatic hyperuricaemia (up to 9 mg/100 ml) only dietary treatment is recommended. In patients with serum urate levels above 9 mg/100 ml or with clinical manifestations of hyperuricaemia drug therapy should be initiated as well. During the first months of drug treatment a low dose of colchicine should be administered concomitantly. Several studies have revealed that this colchicine prophylaxis reduces the potential of developing acute gouty arthritis (Gutman and Yü 1952; Yü and Gutman 1961).

Allopurinol

Allopurinol decreases uric acid formation mainly by inhibiting xanthine oxidase, the enzyme which is responsible for the conversion of hypoxanthine to xanthine and xanthine to uric acid. After the administration of allopurinol uric acid and urinary uric acid excretion fall, accompanied by an increase of urinary oxypurine excretion (Fig. 2). In most patients the replacement of urinary uric acid by hypoxanthine and xanthine is less than stoichiometric (Fig. 2; Rundles et al. 1963; Zöllner and Gröbner 1970). The purine deficit ranges from 10%–30% during a purine-free diet to up to

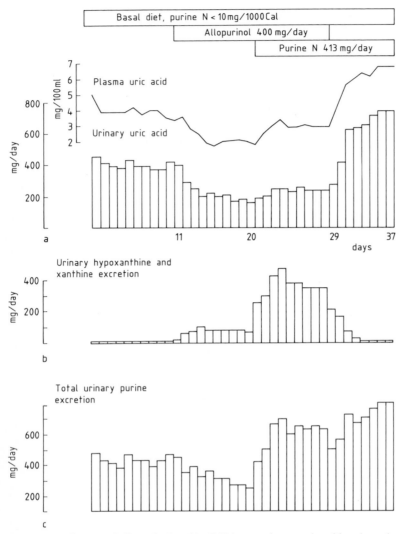

Fig. 2a–c. Influence of allopurinol and/or RNA on **a** plasma uric acid and renal excretion of uric acid, **b** oxypurines and **c** total purines (uric acid + hypoxanthine + xanthine). (From Zöllner and Gröbner 1970)

about 60% during purine ingestion (Zöllner and Gröbner 1970; Löffler and Gröbner 1988). This observation led to the assumption that allopurinol also influences de novo purine synthesis. This effect of allopurinol, which requires normal hypoxanthine-guanine phosphoribosyltransferase activity (HGPRT), can be explained by several mechanisms (Fig. 3):

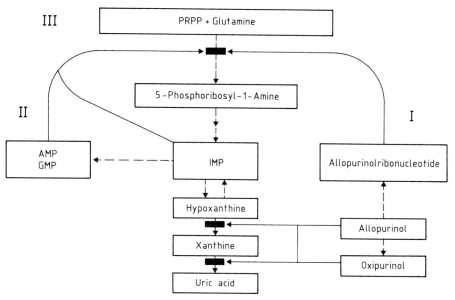

Fig. 3. Possible mechanisms of the influence of allopurinol on purine synthesis de novo. *PRPP*, 5-phosphoribosyl-1-pyrophosphate; *IMP*, inosinic acid; *AMP*, adenylic acid; *GMP*, guanylic acid. (From Zöllner and Gröbner 1987)

1. Allopurinolribonucleotide is an inhibitor of glutamine phosphoribosyl-pyrophosphate-amidotransferase, the rate-limiting enzyme of purine synthesis.
2. The administration of allopurinol leads to an enhanced conversion of hypoxanthine to inosinic acid (IMP). IMP and the purine ribonucleotides derived from IMP are allosteric inhibitors of glutamine phosphoribosylpyrophosphate-amidotransferase.
3. The administration of allopurinol leads to a substantial reduction of erythrocyte phosphoribosylpyrophosphate (PRPP) content (Fox et al. 1970). This effect is due to the consumption of PRPP resulting from the conversion of allopurinol to its ribonucleotide. PRPP is a substrate of glutamine phosphoribosylpyrophosphate-amidotransferase; therefore a reduction of intracellular PRPP concentration reduces purine synthesis de novo.

These three mechanisms probably account for allopurinol's influence on purine synthesis de novo. In addition, the purine deficit during allopurinol treatment could also be explained by the inhibition of xanthine oxidase in the small intestine mucosa with partial loss of oxypurines through the gut (Zöllner and Gröbner 1970) or by inhibition of the absorption of dietary purines (Simmonds et al. 1973).

The administration of allopurinol also leads to a striking increase in the renal excretion of orotidine and orotic acid (Fox et al. 1970; Zöllner and Gröbner 1971). Studies in vivo and in cell culture demonstrated that this effect results from inhibition of orotidine-5-phosphate decarboxylase which catalyses a step in the conversion of orotic acid to uridine-5-phosphate (UMP). In addition, allopurinol therapy results in an increase in the specific activity of both orotate phosphoribosyltransferase (OPRT) and orotidine-5-phosphate decarboxylase (ODC) in erythrocytes (Beardmore et al. 1972). This effect is probably related to enzyme stabilisation, rather than to enzyme synthesis or turnover (Gröbner and Kelly 1975). Allopurinol-induced orotaciduria can be diminished by oral administration of RNA, purine and pyrimidine nucleotides and nucleosides, hypoxanthine and to a lesser extent adenine (Gröbner and Zöllner 1983).

The study of allopurinol metabolism shows that allopurinol has a very short biological half-life of only 2–3 h. Some 3%–10% of an administered dose is excreted with a clearance rate approximately equal to the glomerular filtration rate. Most of the allopurinol (45%–65%) is rapidly oxidized to oxipurinol, with a smaller portion being converted to allopurinol ribonucleoside and allopurinol ribonucleotide. Most of the oxipurinol formed is excreted unchanged by the kidney, with a half-life ranging from 13.5 to 28 h. A small portion is metabolized to the 7-N-ribosyloxipurinol (oxipurinol ribonucleoside) and the 1-N-ribosyl oxipurinol derivative as well as to the corresponding ribonucleotide derivatives (Wyngaarden and Kelly 1983). Factors which affect uric acid excretion generally alter oxipurinol excretion in a similar manner.

The metabolism of allopurinol is influenced by dietary purines. After oral administration of hypoxanthine, inosine, adenine, adenosine and adenine-5-phosphate, Reiter et al. (1984) found a strong decrease in allopurinol-1-riboside excretion. Berlinger et al. (1985) observed an effect of dietary proteins on the renal excretion of oxipurinol.

The average dose of allopurinol required to maintain a serum uric acid level below 6.5 mg/100 ml is 200–300 mg/day; it is rarely necessary to increase the dose. In patients with decreased renal function the dose of allopurinol has to be reduced.

Allopurinol interacts with 6-mercaptopurine, azathioprine, coumarins and other compounds. Side effects of allopurinol therapy appear to be rare. A small percentage of patients find it necessary to discontinue the drug. Allopurinol may lead to the development of gastrointestinal intolerance, skin rashes which are sometimes accompanied by fever, and occasionally toxic epidermal necrolysis, alopecia, bone marrow suppression, granulomatous liver disease and vasculitis. Toxic effects tend to occur more often in the presence of renal insufficiency (Wyngaarden and Kelley

1983; Rundles 1985). The development of xanthine crystalluria or lithiasis as a complication of allopurinol therapy has only been observed in patients with Lesch-Nyhan syndrome or patients with partial HGPRT deficiency. Xanthine stones have also been reported in an adult with lymphosarcoma given allopurinol (Band et al. 1970).

Uricosuric Drugs

A large number of drugs with diverse chemical structure decrease serum urate concentration in humans by enhancing the renal excretion of uric acid. The category of compunds includes salicylates, derivatives of benzoic acid (carinamide, probenecid), phenylbutazone and its derivatives, sulfin-pyrazone, cinchophen, zoxazolamine, niridazole, dicoumarol, phenylin-dandione, benzofurane derivatives and others (Gröbner and Zöllner 1976; Löffler 1982). In uricosuric therapy only probenecid (1–3 g/day), sulfin-pyrazone (200–400 mg/day) and benzbromarone (50–100 mg/day) are usually employed. The uricosuric effect of most agents studied has been at-tributed to inhibition of the tubular reabsorption of filtered urate.

At low dosages some agents with an uricosuric effect, like probenecid, can cause retention of uric acid ("paradoxical retention"), which is attrib-uted to inhibition of tubular uric acid secretion. Administration of higher doses produces uricosuria due to additional inhibition of tubular uric acid reabsorption. There is little known about the molecular mechanism of uric acid transport in the kidney or its influence by uricosuric drugs. The com-bined administration of uricosuric drugs can cause potentiating or antag-onistic effects. The uricosuric actions of probenecid, sulfinpyrazone and zoxazolamine are suppressed by salicylates in humans (Seegmiller and Grayzel 1960). The mechanism of these antagonistic effects is still un-known.

When a potent uricosuric drug is given to normal human subjects at an adequate dosage there is a prompt increase in renal excretion of uric acid, accompanied by a fall in plasma urate, but the urinary uric acid excess dis-appears after about 2 days despite continued administration of the drug (Fig. 4). This does not signify acquired refractoriness since the reduction in plasma urate persists and the urate clearance remains above normal. The normal body pool of urate is so limited, however, that it is depleted after about 2 days of substantially increased drainage through the kidneys. Thereafter, a steady state of balance between production and elimination is reestablished, and the rate of urinary excretion of uric acid is determined by the normal rate of uric acid production. In patients with gout, however, the uricosuric response is more sustained because the body pool of urate is expanded (Fig. 4).

W. Gröbner

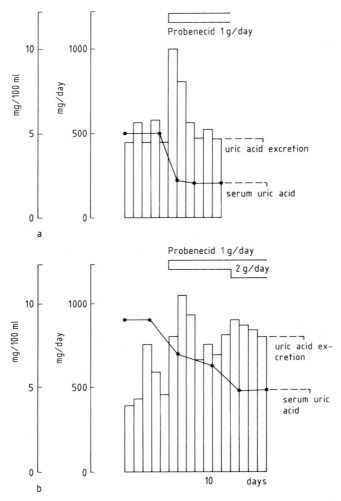

Fig. 4 a, b. Serum uric acid in (**a**) a healthy person and (**b**) in a patient with gout after daily administration of probenecid. (From Zöllner and Gröbner 1976)

The enhanced renal excretion of uric acid can cause uric acid precipitation in the kidney. This complication can be minimized by initiating uricosuric therapy with a low dose and increasing it over a period of days or weeks, maintaining an adequate urine flow and neutralizing the urine with orally administered sodium bicarbonate or sodium citrate during the early course of drug therapy. After hyperuricaemia has been controlled and the urinary uric acid has returned to normal, these precautions are no longer necessary in most patients. The frequency of side effects of probe-

Table 1. The influence of allopurinol (100 or 300 mg/day), benzbromarone (20 mg/day) or a fixed combination of 20 mg benzbromarone and 100 mg allopurinol on serum uric acid concentration; the experiments were performed under standardized nutrition conditions. (From Löffler et al. 1983)

	Allopu-rinol 100 mg ($n=6$)	Allopu-rinol 300 mg ($n=8$)	Benz-bromarone 20 mg ($n=8$)	20 mg Benz-bromarone + 100 mg allopurinol ($n=8$)
Control (mg/dl)	6.8 ± 0.8	7.2 ± 1.1	6.6 ± 1.2	6.9 ± 1.2
With medication (mg/dl)	5.1 ± 0.9	3.9 ± 1.1	4.3 ± 1.3	4.1 ± 1.1
Change (%)	-25	-46	-35	-41

necid, sulfinpyrazone and benzbromarone is low. Hypersensitivity reactions and gastrointestinal disturbances have been observed.

Combined Use of Allopurinol and a Uricosuric Agent

Combined therapy with allopurinol and a uricosuric agent may be particularly effective in patients with extensive tophaceous deposits where substantial amounts of preformed urate must be eliminated. Uricosuric drugs, however, increase the clearance of oxipurinol in humans and thus diminish the degree of xanthine oxidase inhibition (Elion 1966; Elion et al. 1968).

In the Federal Republic of Germany a combination of 100 mg allopurinol and 20 mg benzbromarone was introduced several years ago. The effect of this fixed combination is the same as that of 300 mg allopurinol and less than that of 100 mg benzbromarone (Table 1; Löffler et al. 1983). The assumption that this combination involves a lower therapeutical risk than an equally effective monotherapy, especially allopurinol therapy, has not been adequately proven so far. In addition, the frequency of side effects which are not dose dependent is presumably increased with combined therapy.

References

Band PR, Silverberg DS, Henderson JF, Ulan RA, Wensel RH, Banerjee TK, Little AS (1970) Xanthine nephropathy in a patient with lymphosarcoma treated with allopurinol. N Engl J Med 283:354

Beardmore TD, Cashman JS, Kelley WN (1972) Mechanism of allopurinol mediated increase in enzyme activity in man. J Clin Invest 51:1823

Berlinger WG, Park GD, Spector R (1985) The effect of dietary protein on the clearance of allopurinol and oxipurinol. N Engl J Med 313:771

Colling M, Wolfram G (1987) Zum Einfluß des Garens auf den Puringehalt von Lebensmitteln. Z Ernährungswiss 26:214

Elion GB (1966) Enzymatic and metabolic studies with allopurinol. Ann Rheum Dis 25:608

Elion GB, Yü TF, Gutman AB, Hitchings GH (1968) Renal clearance of oxipurinol, the chief metabolite of allopurinol. Am J Med 45:69

Fox IJ, Wyngaarden JB, Kelley WN (1970) Depletion of erythrocyte phosphoribosylpyrophosphate in man, a newly observed effect of allopurinol. N Engl J Med 283:1177

Fox RM, Royse-Smith D, O'Sullivan WJ (1970) Orotidinuria induced by allopurinol. Science 168:861

Gröbner W, Kelley WN (1975) Effect of allopurinol and its metabolic derivatives on the configuration of human orotate phosphoribosyltransferase and orotidine-5-phosphate decarboxylase. Biochem Pharmacol 24:379

Gröbner W, Zöllner N (1976) Uricosurica. In: Zöllner N, Gröbner W (eds) Gicht. Springer, Berlin Heidelberg New York, p 491, (Handbuch der Inneren Medizin, vol 7/3)

Gröbner W, Zöllner N (1983) Der Einfluß von Nahrungspurinen und -pyrimidinen auf die Pyrimidinsynthese des Menschen. Klin Wochenschr 61:1191

Gutman AB, Yü TF (1952) Gout, a derangement of purine metabolism. Adv Intern Med 5:227

Löffler W (1982) Urikosurika. In: Zöllner N (ed) Therapie und Prognose von Hyperurikämie und Gicht. Springer, Berlin Heidelberg New York, p 73 (Hyperurikämie und Gicht, vol 5)

Löffler W, Gröbner W (1988) A study of dose-response relationships of allopurinol in the presence of low or high purine turnover. Klin Wochenschr 66:153

Löffler W, Gröbner W, Zöllner N (1983) Harnsäuresenkende Wirkung einer Kombination von Benzbromaron und Allopurinol-Untersuchungen unter standardisierten Ernährungsbedingungen. Arzneimittelforschung 33(II):1687

Reiter S, Löffler W, Gröbner W, Zöllner N (1984) Influence of dietary purines on the metabolism of allopurinol in man. World Rev Nutr Diet 43:187

Rundles RW (1985) The development of allopurinol. Arch Intern Med 145:1492

Rundles RW, Wyngaarden JB, Hitchings GH, Elion GB, Silberman HR (1963) Effects of xanthine oxidase inhibitor on thiopurine metabolism, hyperuricemia and gout. Trans Assoc Am Physicians 76:126

Seegmiller JE, Grayzel AJ (1960) Use of the newer uricosuric agents in the management of gout. JAMA 173:1076

Simmonds HA, Rising TJ, Cadenhead A, Hatfield PJ, Jones A, Cameron JS (1973) Radioisotope studies of purine metabolism during administration of guanine and allopurinol in the pig. Biochem Pharmacol 22:2553

Wyngaarden JB, Kelley WN (1983) "Gout". In: Stanbury JB, Wyngaarden JB, Fredrickson DS, Goldstein JL, Brown MS (eds) The metabolic basis of inherited disease, 5th edn. Mc Graw-Hill, New York, p 1043

Yü TF, Gutman AB (1961) Efficacy of colchicine prophylaxis in gout. Ann Intern Med 55:179

Zöllner N (1975) Diet and gout. Nutrition 1:267

Zöllner N, Gröbner W (1970) Der unterschiedliche Einfluß von Allopurinol auf die endogene und exogene Uratquote. Eur J Clin Pharmacol 3:56

Zöllner N, Gröbner W (1971) Influence of oral ribonucleic acid on orotaciduria due to allopurinol administration. Z Gesamte Exp Med 156:317

Zöllner N, Gröbner W (1987) Purinstoffwechsel; Urikosurika, Urikostatika. Pharmakotherapie der Gicht. In: Forth W, Henschler D, Rummel W (eds) Allgemeine und spezielle Pharmakologie und Toxikologie. Wissenschaftsverlag, B. J., p 362

Zöllner N, Griebsch A, Gröbner W (1972) Einfluß verschiedener Purine auf den Harnsäurestoffwechsel. Ernährungsumschau 3:79

The Development of Antiviral Compounds Against HIV and Herpesviruses

K. L. POWELL

The Discovery of Antiviral Drugs

The problem of discovering a new drug of any type is made relatively simple once a lead compound has been obtained. Much time is given to the discovery of such leads, for example, in screening for activity of drugs against viruses. Given that such a screen is available, the immediate question is which compounds to test in such a screen (we may make the assumption that between 10^5 and 10^6 compounds are potential candidates). In the area of antiviral drugs the logical choice is to screen low molecular weight substrate analogues of the enzymes involved in the essential replicative functions of the virus nucleic acids. It is not surprising, therefore, that the antiviral drugs that are in widespread use are predominantly nucleoside analogues. Drug discovery in the antiviral area is moving rapidly from the blind approach of screening nucleosides for activity to the specific design of new agents based on more detailed information of virus enzymes, but it is to be expected that in the short term purines and pyrimidines will remain vitally important to the clinical treatment of virus disease.

The Early Years: Non-Specific Agents

From the mid 1950s to the mid 1970s the antiviral agents reported were confined to those with limited specificity or therapeutic utility. Probably the most successful were the 5-substituted pyrimidines such as iododeoxyuridine or trifluorothymidine, which have been widely used in the treatment of herpetic keratitis where their cytotoxicity is of less importance. The wide spectrum of pyrimidines and purines which have been made and tested as antiviral agents has been reviewed in depth by De Clercq (1984, 1987), and in this discussion I will limit my remarks to those drugs which have found widespread clinical use.

155

Antiherpesvirus Drugs

The useful effects of the early non-specific compounds have been eclipsed by the discovery of acyclovir (Zovirax). This compound was discovered in the mid 1970s by researchers at Burroughs Wellcome (Elion 1986). These researchers had been stimulated by the discovery of the antiviral activity of adenine arabinoside to look for antiviral activity in the purines. It was particularly fortuitous that at the same time antiviral screening was available to these chemists. One series of compounds examined were purine analogues which had been altered in the ribose moiety to test their ability to interact with adenosine deaminase. The acyclic derivative of guanine (acyclovir) was found to be the most active of these compounds and has, of course, become a highly successful drug. Acyclovir is specific because it is recognised by the virus thymidine kinase and phosphorylated by it, but not by the equivalent host cell enzyme (Fyfe et al. 1978). Once the drug has been phosphorylated to the monophosphate, conversion to the di- and tri-phosphate is achieved by cellular enzymes (Miller and Miller 1980, 1982). The triphosphate is then better recognised as a substrate by the viral than by the cellular DNA polymerases. Reaction of the viral DNA polymerase results in both termination of the growing DNA chain (due to the lack of the 3'-hydroxyl group in acyclovir) and inactivation of the DNA polymerase (Furman et al. 1979, 1984). These stages in the mechanism of acyclovir lead it to have a 3000-fold difference in its IC_{50} for Vero cell growth and for inhibition of virus (Elion 1986), making it a very specific compound.

Acyclovir has been and continues to be a very successful drug, both commercially and in the clinic. This success has not brought as yet any major problems of resistance to the drug, and yet each type of drug resistance seen in laboratory studies of virus mutants (Larder and Darby 1984) have found clinical counterparts. These mutants have three phenotypes: they either are thymidine-kinase negative, have a thymidine kinase with altered substrate specificity, or have an altered DNA polymerase. It is apparent that mutants with such altered proteins must have reduced ability to replicate and spread their infection through a patient population.

There have been many attempts to improve on acyclovir as an antiherpesvirus drug. Of the many acyclovir analogues, only the hydroxymethyl 3' carbon-substituted form (ganciclovir) has made much progress. This compound has the advantage of better activity against human cytomegalovirus (CMV) than acyclovir. Unfortunately, toxicity has limited the use of ganciclovir, although it is a drug which has been used to treat CMV disease in AIDS patients. There are still many laboratories working on the analogues of acyclovir and these may well yield an interesting agent, but it should be remembered that acyclovir's combination of selectivity, potency and oral availability is very difficult to surpass.

Azidothymidine

The causative agent of AIDS, the human immunedeficiency virus (HIV), was identified early (Barre-Sinoussi et al. 1983). The attention of virologists has been concentrated on the various gene products of this virus which might act as targets for antiviral drugs. The virus belongs to the retrovirus group and has an RNA genome. A feature of the replication of these viruses is the conversion of their genomes to double-stranded DNA by a virus-encoded reverse transcriptase (RT) present within the virus particles. This enzyme has been studied extensively in animal RNA tumour viruses. Several inhibitors of the enzyme that were discovered were known prior to the advent of AIDS. The most interesting of these is azidothymidine (AZT), which is now being widely used as an anti-AIDS drug (Mitsuya et al. 1985). AZT is closely related to thymidine and is currently made using this natural substance as starting material.

Clinical Experience

The results of clinical trials with AZT are well known and I will just touch on them here (Yarchoan et al. 1986; Fischl et al. 1987). The drug was found to be easily administered; it has good oral bioavailability (about 60%). AZT has a relatively short half-life; virus inhibitory levels are easily obtained in patients. The chief route of drug clearance is via an inactive metabolite. Importantly, the drug is efficiently transported across the blood-brain barrier and would thus be expected to inhibit the virus in clinical cases of AIDS virus brain infection. One interesting aspect of antiviral drugs is that we do not yet know the concentration of the drug or the duration of its application that is required for clinical effect. Is it necessary to administer the drug at its IC_{50} level continuously or only, for example, once a day? Only clinical experience can give us this information.

In the phase one study the drug was found to bring about improvement in several parameters (both clinical and immunological) whilst having haematological toxicity problems which could be tolerated. Thus, a phase two study was initiated with 281 patients. Initially this was a randomised, double-blind, placebo-controlled study, but on examination of the preliminary results the trial was opened and patients receiving placebo were placed on the drug. The chief reason for this was that only 1 patient receiving Retrovir (AZT) had died, as compared with 19 in the placebo group. Not only did the drug reduce mortality, but it also significantly reduced the incidence of opportunistic infection. Positive indications were also seen in the quality of the patient's life, patient's weight gain, T_4 cell counts, delayed-type hypersensitivity (DTH) skin test response and, virologically, in P24 antigen levels. Further clinical experience with the drug has largely con-

157

firmed this early experience. Unfortunately, no other anti-AIDS drug has yet been discovered which appears as promising as Retrovir. Retrovir clearly has some side effects and these have been reviewed recently in some detail (Fischl et al. 1987).

Mode of Action

AZT is phosphorylated in infected or uninfected cells by the same enzymes that are responsible for the phosphorylation of thymidine (Furman et al. 1986). The drug is a good substrate for cellular thymidine kinase. On phosphorylation to the monophosphate it is bound strongly by the cell thymidylate kinase, but is converted slowly by this enzyme to the diphosphate. Thus, the monophosphate may be considered a substrate-based inhibitor of the cell thymidine monophosphate (TMP) kinase. Intracellular levels of AZT monophosphate have been recorded as high as 800 μM in cells treated with 50 μM AZT. This level of the monophosphate greatly exceeds the level of the triphosphate, which is the level at which specificity occurs. Inhibition of the AIDS virus reverse transcriptase (RT) occurs at very low levels of AZT triphosphate. Using cloned HIV RT we have measured the IC_{50} of the triphosphate on the enzyme as <0.1 μM (Larder et al. 1987 b). Although the drug does have inhibitory effects on cell enzymes, these occur at a level hundreds of times higher than that required to inhibit the viral enzyme.

Future Antiviral Agents

Our work at Wellcome is now aimed at developing new antiviral agents. Despite the success of current drugs based on pyrimidine and purine analogues, our rationale is that detailed studies of particular virus proteins will enable us to make more novel virus inhibitors, using the structure of the virus proteins as a guide. The success of AZT as a treatment for AIDS has encouraged us to concentrate on the virus RT as a target and the rest of my paper concerns these studies.

Large-Scale Production of HIVRT

Purification of HIV RT from virus particles allows limited characterisation of the HIV enzyme, but does not produce enough enzyme for crystallographic studies. In order to do these studies large amounts of enzyme were required which could only be obtained by gene cloning. Our first step was to acquire part of the AIDS genome as cloned material from Dr Malcolm Martin at NIH. The first task was to "tailor" these genes to allow expression of the RT. The virus *pol* gene encodes three proteins – a protease, the RT and an endonuclease – which are made as a polyprotein in virus-infected cells. Cleavage of the polyprotein by proteases yields each gene product (Muesing et al. 1985; Ratner et al. 1985; Sanchez-Pescador et al. 1985; Wain-Hobson et al. 1985). We wanted to express the RT gene in the absence of the other products. Thus, from the original clone a construct was made in the M13 phage system (which allows easy genetic manipulation) by directly cloning the *Bg*II-*Eco*RI fragment containing the protease and RT encoding portions and part of the endonuclease into mp*tac*18, a specially constructed cloning vector (B. Larder, unpublished information) designed to contain the RT gene behind the strongly inducible *tac* promoter. We next removed the extraneous material by a series of specific deletions. Firstly, the endonuclease remnants were removed by oligonucleotide-directed mutagenesis, using a single "deletion" oligonucleotide with 15 nucleotides complementary to the C-terminal portion of the RT and 15 complementary to the 5′ end of the *lacZ* gene in mp*tac*18. The protease region was removed from the resulting clone by first inserting an *Eco*R1 restriction enzyme site just upstream from the RT start by site-directed mutagenesis, then removing the DNA from between the pair of *Eco*R1 sites thus created, forming the phage mpRT4. The RT produced by each of these clones was examined and each reduction in the material superfluous to the RT was found to improve the level of RT activity.

These manipulations had given us levels of RT activity which were useful and allowed detailed biochemical study, but to get sufficient protein for crystallisation purposes still higher levels of enzyme were required. These levels wre achieved by removing the "cassette" we had created from mpRT4 and inserting it into the high-level expression plasmid pKK233-2. This resulted in the plasmid pRT1, which on insertion into *E. coli* produced the RT polypeptide at the level of about 10% of the *E. coli* total cell protein. This level of protein production was mirrored by enzyme activity and represented authentic enzyme as judged by its reaction with AIDS patient sera, thus allowing purification of the RT in quantity (Larder et al. 1987 b).

Crystallisation of HIV RT

Using the high-level expression system defined above we have obtained more than 0,5 g of RT. This material has been purified using a monoclonal antibody affinity column which allows large-scale purification of the enzyme in a single step (Tisdale et al. 1988). This enzyme is almost completely composed of a 66 kDa protein which is contaminated with minor amounts of a smaller polypeptide. The protein has been used in crystallisation trials and has been shown to produce rod-shaped crystals. These crystals were analysed by gel electrophoresis and were shown to be composed of equal amounts of two polypeptides of 66 kDa and 51 kDa (Lowe et al. 1988). These are very similar to the polypeptides found in virus particles and pose the question: what is the composition of HIV RT?

Our experiments demonstrate that the HIV RT of 66 kDa or 66 and 51 kDa proteins is a hetero- or homodimer. The 66 kDa homodimer can be cleaved by a non-specific protease to produce the heterodimer. Both forms of the enzyme are active and produce similar results when the enzyme inhibitor AZT triphosphate is added (Lowe et al. 1988). These results suggest that the enzyme can be active in a multitude of forms, which creates problems for the crystallographer who wants a homogenous preparation. Our efforts in this direction have produced three further crystal forms thus far and may produce more as we work towards crystals capable of producing good resolution.

The three-dimensional structure of the enzyme will not of course reveal the importance of individual amino acids in the active sites of the enzyme. To begin to come to terms with this difficulty we have looked at sequence similarities betwen the HIV RT gene and that of closely related polymerase genes revealed by Johnson et al. (1986). We have then mutated each of these regions of the gene to produce site-specific mutants in the conserved areas. Without going into this study (Larder et al. 1987a) in detail I would like to point out how interesting the results can be. Firstly, to convince you that the result of the mutation is not always interesting, the change from pro to leu in mutant RTM11 in conserved region D had no effect on the enzyme. Further mutants show much more interesting effects; for example, in region E, changing one of a pair of conserved aspartates to histidine totally abolished activity, thus identifying a region of great interest in the enzyme's activity. Two other mutants in this region, RTM2 and RTM4, also reduced enzyme activity, confirming this result. Secondly, two mutants in region B RTM5 and RTM6 drastically altered the enzyme – it was less active than the wild type and was markedly less sensitive to inhibition by AZT triphosphate and phosphonoformic acid. These changes are most interesting, indicating both where the triphosphate binding site of the enzyme might be located and raising the possibility of alterations in sensitivity of AIDS virus to drugs by alteration of the RT (of course, we do not know if a virus carrying such a mutation would be viable or if it would be able to induce full-blown AIDS).

Conclusions

Our studies are progressing towards the determination of the structure of the HIV RT. We already know some of the critical amino acids which are essential for enzyme activity. Given this information we hope to be able to design new agents which are even more potent and successful than current nucleoside analogues. Such work is, of course, very dependent on the contributions of many scientists in the field of purine and pyrimidine metabolism and is indeed very relevant to the subject of this symposium.

Acknowledgements. The work described in this paper on HIV reverse transcriptase has been done by C. Bradley, D. Lowe, D. Stammets, K. Powell, D. Purifoy, G. Darby, B. Larder, M. Tisdale, S. Kemp and P. Ertl. I thank them and also M. Parkinsen for her help with this manuscript.

References

Barre-Sinoussi F, Cherman JC, Rey F, Nugeybe MT, Charmaret S, Gruest J, Dauguet C, Axler-Blin C, Vezinet-Brun F, Rouzioux C, Rozenbaum W, Montagnier L (1983) Isolation of a T-lymphotropic retrovirus from a patient at risk of acquired immune deficiency syndrome (AIDS). Science 220:868–870

De Clercq E (1984) Biochemical aspects of the selective antiherpes activity of nucleoside analogues. Biochem Pharmacol 33:2159–2173

De Clercq E (1987) Molecular targets for selective antiviral chemotherapy. A multidisciplinary approach. In: De Clercq E, Walker RT (eds) Antiviral drug development. Plenum, New York, pp 97–122

Elion G (1986) History, mechanism of action, spectrum and selectivity of nucleoside analogs. In: Mills C, Corey L (eds) Antiviral chemotherapy: new directions for clinical applications and research. Elsevier, New York, 118–137

Fischl MA, Richman DD, Grieco MH, Gottlieb MS, Volberding PA, Laskin OS, Leedom JM, Groopman JE, Mildvan D, Schooley RT, Jackson GG, Durack DT, Phil D, King D, the AZT Collaborative Working Group (1987) The efficacy of Azidothymidine (AZT) in the treatment of patients with AIDS and AIDS-related complex. A double-blind placebo-controlled trial. N Engl J Med 317:185–191

Furman PA, St Clair MH, Fyfe JA, Rideout JL, Keller PM, Elion G (1979) Inhibition of herpes simplex virus induced DNA polymerase and viral DNA replication by 9-[2-hydroxyethoxymethyl]-guanine and its triphosphate. J Virol 32:72–77

Furman PA, St. Clair MH, Spector T (1984) Acyclovir triphosphate is a suicide inactivator of the herpes simplex virus DNA polymerase. J Biol Chem 259:9575–9579

Furman PA, Fyfe JA, St Clair MH, Weinhold K, Rideout JL, Freeman GA, Lehrman SN, Bolognesi DP, Broder S, Mitsuya H, Barry DW (1986) Phosphorylation of 3'-Azido-3'-deoxythymidine and selective interaction of the 5'-triphosphate with human immunodeficiency virus reverse transcriptase. Proc Natl Acad Sci USA 83:8333–8337

Fyfe JA, Keller PA, Furman PA, Miller RL, Elion GB (1978) Thymidine kinase from herpes simplex virus phosphorylates the new antiviral compound 9-[2-hydroxyethoxymethoxy]-guanine. J Biol Chem 253:8721–8727

Johnson MS, McLure MA, Feng D-F, Gray J, Doolittle RF (1986) Computer analysis of retroviral pol genes: assignment of enzymatic functions to specific sequences and homologies with non-viral enzymes. Proc Natl Acad Sci USA 83:7648–7652

Larder BA, Darby GK (1984) Virus drug-resistance: mechanisms and consequences. Antiviral Res 4:1–42

Larder BA, Purifoy DJM, Powell KL, Darby GK (1987a) Site-specific mutagenesis of AIDS virus reverse transcriptase. Nature 327:716–717

Larder BA, Purifoy DJM, Powell KL, Darby G (1987b) AIDS virus reverse transcriptase defined by high level expression in *E. coli*. EMBO J 6:3133–3137

Lowe DM, Aitken A, Bradley C, Darby G, Larder BA, Powell KL, Purifoy DJM, Tisdale M, Stammers DK (1988) HIV reverse transcriptase: crystallization and analysis of domain structure by limited proteolysis. Biochemistry (In Press)

Miller WH, Miller RL (1980) Phosphorylation of acyclovir monophosphate by GMP kinase. J Biol Chem 253:7204–7207

Miller WH, Miller RL (1982) Phosphorylation of acyclovir diphosphate by cellular enzymes. Biochem Pharmacol 31:3879–3884

Mitsuya H, Weinhold KJ, Furman PA, St. Clair M, Lehrman SN, Gallo RC, Bolognesi D, Barry DM, Broder S (1985) 3'Azido-3'deoxythymidine (BW A509U): an antiviral agent that inhibits the infectivity and cytopathic effect of human T-lymphotropic virus type III/lymphoadenopathy associated virus *in vitro*. Proc Natl Acad Sci USA 82:7096–7100

Muesing MA, Smith DH, Cabradilla CD, Benton CV, Lasky LA, Capon DJ (1985) Nucleic acid structure and expression of the human AIDS/lymphoadenopathy retrovirus. Nature 313:450–458

Ratner L, Haseltine W, Patarca R, Livak KJ, Starchich B, Josephs SF, Doran ER, Rafalski JA, Whitehorn EA, Baumeister K, Ivanoff L, Petteway SR, Pearson ML, Lautenberger JA, Papas TS, Ghrayeb J, Chang NT, Gallo RC, Wong-Staal F (1985) Complete nucleoside sequence of the AIDS virus, HTLV-III. Nature 313:277–284

Sanchez-Pescador R, Power MD, Barr PJ, Steimer KS, Stempien MM, Brown-Shimer SL, Gee WW, Renard A, Randolph A, Levy JA, Dina D, Luciw PA (1985) Nucleotide sequence and expression of an AIDS-associated retrovirus (ARV-2). Science 227:484–492

Tisdale M, Larder BA, Lowe DM, Stammers DK, Purifoy DJM, Ertl PF, Bradley C, Kemp S, Darby GK, Powell KL (1988) Characterization of HIV RT using monoclonal antibodies: the role of the C terminus in antibody reactivity and enzyme function. J Virol (In press)

Wain-Hobson S, Sonigo P, Danos O, Cole S, Alizon M (1985) Nucleotide sequence of the AIDS virus, LAV. Cell 40:9–17

Yarchoan R, Weinhold KJ, Lyerly HK, Gelmann E, Blum RM, Shearer GM, Mitsuya H, Collins JM, Myers CE, Klecker RW, Markham PD, Durack DT, Lehrman SN, Barry DW, Fischl MA, Gallo RC, Bolognesi DP, Broder S (1986) Administration of 3'-Azido-3'-deoxythymidine, an inhibitor of HTLV-III/LAV replication, to patients with AIDS or AIDS-related complex. Lancet I:575–580

C. M. Feek, Wellington, New Zealand; C. R. W. Edwards, Edinburgh, UK

Endocrine and Metabolic Disease

With a contribution by A. D. Struthers

1988. 43 figures. Approx. 120 pages. (Treatment in Clinical Medicine). Soft cover
ISBN 3-540-19504-1

Contents: Systematic Review of Endocrine Diseases. – Clinical Pharmacology of Endocrine Drugs. – Subject Index.

Endocrine and Metabolic Disease is in two sections. The first is a systematic review of clinical and therapeutic aspects of endocrine diseases, whilst the second considers the clinical pharmacology of drugs used in endocrinology. All the major areas of endocrinology and metabolism are covered. Thyroid, adrenal and pancreatic disease are discussed in depth and there are separate chapters on the ovary and testes as well as on the pituitary gland and parathyroids. There is a useful pharmacopoeia of drugs used in endocrinology at the end of the volume.

Springer-Verlag Berlin
Heidelberg New York London
Paris Tokyo Hong Kong

Springer

F. Krück, Bonn; K. Thurau, Munich, FRG (Eds.)

Endocrine Regulation of Electrolyte Balance

1986. 71 figures, 18 tables. XI, 150 pages. Soft cover. ISBN 3-540-16837-0

Springer-Verlag Berlin
Heidelberg New York London
Paris Tokyo Hong Kong

Springer